8	more than one	more than one	mean score and proportion	national sample survey data from the Southeast Regional Survey Project, University of North Carolina	tests hypotheses about functions based on table marginals	multivariate analysis of variance (MANOVA)
9	more than one	more than one	rank correlation coefficient gamma	sample survey data from the Houston Community Study	demonstrates use of complex (compound) function of cell frequencies	partial correlation analysis
10	more than one	one	mean score based on ranked data	sample survey data from the Houston Community Study	demonstrates multivariate analysis of ranked data	multivariate analysis of variance (MANOVA)
11	one	more than one	survival curve function	data on the use of services provided by the Kaiser Health Plan, Portland, Oregon	demonstrates use of survival data in an HMO setting using complex function	follow-up life table analysis

Public
Program Analysis
A NEW CATEGORICAL DATA APPROACH

Public Program Analysis
A NEW CATEGORICAL DATA APPROACH

Ron N. Forthofer
Robert G. Lehnen

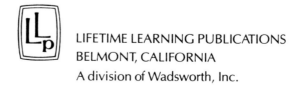
LIFETIME LEARNING PUBLICATIONS
BELMONT, CALIFORNIA
A division of Wadsworth, Inc.

Designer: Rick Chafian

Developmental Editor: Don Yoder

Printed in the United States of America

1 2 3 4 5 6 7 8 9 10—85 84 83 82 81

Library of Congress Cataloging in Publication Data

Forthofer, Ron N., 1944-
    Public program analysis.

    Bibliography: p.
    Includes index.

    1. Evaluation research (Social action programs)
—Statistical Methods.   2. Public health—Evalua-
tion—Statistical methods.   3. Least squares.
4. Multivariate analysis.   I. Lehnen, Robert G.
II. Title.
HA29.F575   361.6'1'0724   81-1125
ISBN 0-534-97974-2   AACR2

To GARY G. KOCH
Teacher, Colleague, and Friend

Preface

The past two decades have witnessed a dramatic increase in the demand for analyses of health and public affairs program data. Governments at the federal and state levels have stimulated, if not mandated, much of this increased demand and spawned numerous not-for-profit and quasi-public organizations—research groups, lobbying organizations, consulting firms—who actively use analytic techniques.

Program analysis is a broad term that encompasses activities such as program planning and prediction, program assessment, and program evaluation. Though purposes differ and techniques vary, the common denominator for all applications is the attempt to use quantitative techniques to provide comprehensive and objective analyses. Program analysis in public health and public affairs shares another common feature: In many cases the data collected for these applications are *categorical* in nature—that is, discrete information represented by categories. Whether a defendant in a criminal trial is convicted (yes, no), the number of times a person visits a community health center in a month (0, 1, 2, 3, . . .), how a person feels about a proposed program (agree, disagree, not sure, no opinion)—these are examples of categorical data.

This book describes a multivariate categorical data analysis technique—the weighted-least-squares (WLS) approach developed by Grizzle, Starmer, and Koch (GSK)—applied to program analysis in health and public affairs. It is written for in-service professionals who desire an introduction to applied categorical data analysis and for preservice students who are studying quantitative methods.

The WLS approach allows practitioners to analyze a wide range of problems, whereas previously they had to learn a different technique, if one existed, for each new type of problem encountered. Moreover, the WLS approach produces estimates that are "best asymptotic normal," which means that with large samples there is no better approach for obtaining statistical estimates.

Our emphasis is on the application of the WLS approach to problems in public health and public affairs. Some of the material is directed toward the statistically experienced reader, but the book is also suited for those who have knowledge only of the analysis of variance. To aid the less experienced reader, Appendix A contains an introduction to matrices. Appendix B, which demonstrates the use of matrices and the linear model in analysis of variance situations, is strongly recommended as preliminary reading for the statistically inexperienced reader.

We have not attempted to describe every technique for program analysis nor to discuss all the available methods for analyzing categorical data. Rather we have focused on the description of one general and powerful approach: the WLS method. We demonstrate the flexibility of this procedure in the analysis of a variety of categorical data problems, but this book is not a complete catalog of all possible applications of this methodology. The tables inside the front cover summarize the applications covered in this book. Because the WLS methodology is too broad to be exhausted in one introductory book, Chapter 12 reviews advanced WLS topics and applications.

By the end of this book the reader will be better able to analyze and interpret multivariate categorical data evolving from situations similar to those presented here. Moreover, he or she will have a better understanding of the statistical basis for the conclusions reached. For more complex problems the reader will be able to communicate more effectively with the statistician for consultation and assistance.

One person—Professor Gary G. Koch of the University of North Carolina —has indirectly played a central role in the preparation of this book. He has generously given us his ideas, comments, and criticisms and has patiently instructed us on many aspects of categorical data analysis. We owe him a great personal and professional debt. Although we have judiciously considered many points of view, the final decision regarding this book's organization, content, and philosophy are strictly our own and should not be attributed to Professor Koch or others.

Ron Forthofer has a special debt to Professor James E. Grizzle of the University of North Carolina, who provided him his first exposure to categorical data analysis.

We also wish to acknowledge the following people for their efforts and contributions: Mary Forthofer, Paul Gregor, Arthur Badgett, Jon-Ing Lin, Robert Hardy, Jay Glasser, Peter Imrey, and Richard Landis. Gay Robertson and her staff deserve special thanks for producing the many typed drafts.

Ron N. Forthofer

Robert G. Lehnen

Contents

Preface   vii

PART I: FOUNDATIONS

1. Multidimensional Contingency Tables   3

2. Guidelines for Analysis   7

    What Are Categorical Data?   7
    Is This a Categorical Data Problem?   9
    What Function Should Be Analyzed?   9
    What Sample Size Is Required?   12
    Are Zero Cells a Problem?   14
    How Many Analyses Should Be Performed?   15
    What Levels of Significance Should Be used?   16
    Summary   17

3. Contingency Table Analysis: The WLS Approach   19

    The Contingency Table   19
    Formation of Functions   22
    The Variance of a Function   24

ix

The Linear Model and Categorical Data   26
An Example of WLS Analysis   27
Additive Association   30
Multiplicative Association   31
Summary   35
Exercises   36

PART II: SIMPLE APPLICATIONS OF THE WLS APPROACH

4. One Response and Two Factor Variables   41

Analyzing Sentences in Criminal Court   41
Defining the Variables   42
The Response Function   43
Specifying the Model   44
Estimating Parameters   45
Evaluating the Model's Fit   46
Testing Hypotheses   48
Interpreting the Model   48
Summary   49
Exercises   50
Appendix: GENCAT Input   54

5. Interaction Among Factor Variables   55

Adding Complexity   55
Modeling Interaction   57
A Modular Model   59
Comparison of Models   65
Summary   66
Exercises   67
Appendix: GENCAT Input   73

6. Mean Scores   75

Measuring HMO Usage   75
Formation of Means   79
The Prediction Model   81
Defining New Factor Variables   85
A Reduced Model   88
Using the Model   91
Summary   92
Exercises   93
Appendix: GENCAT Input   95

7. Log-Linear Models   98

Analyzing Air Pollution Effects   98
The Data   99
Defining the Variables   100
A Rationale for the Logit   101
The Logit Model   106
Comparison of WLS and ML Results   107
An Odds-Ratio Interpretation   108
Lead—A Significant Factor?   109
Summary   110
Exercises   111
Appendix: GENCAT Input   118

PART III: ADVANCED APPLICATIONS OF THE WLS APPROACH

8. Multiple Response Functions   121

Examining Trust in Government   122
Structure of a Multiresponse Table   124
Mean Trust Scores   126
Multiple Proportions   131
Summary   137
Exercises   138
Appendix: GENCAT Input   141

9. Rank Correlation Methods   146

Are People Consistent in Their Responses?   146
A Rank Correlation Coefficient   149
Computing Gamma Using GENCAT   153
Examining Consistency   157
Results: No Education Effect   158
Summary   158
Exercises   159
Appendix: GENCAT Input   162

10. Rank Choice Analysis   164

Choosing Among Competing Alternatives   165
Discrete-Choice Data and Few Choices   166
Discrete-Choice Data and Many Choices   168
Continuous-Choice Data   172
Summary   178
Exercises   179
Appendix: GENCAT Input   186

11. Follow-Up Life Table Analysis   191

    Predicting HMO Membership   191
    The Follow-Up Life Approach   192
    WLS and the Follow-Up Life Table   194
    Specifying the Model   199
    Age and Year of Enrollment   205
    Summary   205
    Exercises   206
    Appendix: GENCAT Input   208

12. Selected WLS Literature   211

    Paired Comparisons   211
    Observer Agreement   212
    Supplemented Margins   212
    Repeated Measures   212
    Partial Association   213
    Competing Risks   213
    Multiple-Record Systems   213
    Complex Sample Surveys   213
    Complex Functions   214
    Functional Asymptotic Regression Methodology (FARM) 214
    Variable Selection   214
    Comparisons to Parametric Regression   215
    Philosophy of Data Analysis   215

APPENDIXES

Appendix A: Matrix Notation   219

    Matrix Definitions   219
    Matrix Arithmetic   221
    The Inverse Matrix   226
    System of Linear Equations—Scalar Presentation   228
    System of Linear Equations—Matrix Presentation   231
    Two Transformations: e and the Natural Logarithms 232
    Operations with e and the Natural Logarithm   234
    Summary   235
    Exercises   236

Appendix B: The Linear Model   238

    Traditional Approach to ANOVA   239
    Linear Model Approach to ANOVA   243

Coding Methods   247
Testing Hypotheses   252
Two-Way ANOVA   254
Interaction   257
Summary   264
Exercises   265

Appendix C: Table of Chi-Square Values   267

Appendix D: The GENCAT Computer Program   269

An Overview of GENCAT   271
Entering the Data to GENCAT   271
Left-Hand Side of the Equation   272
Right-Hand Side of the Equation   272
Testing Individual Hypotheses   272
Summary of Major Input to GENCAT   273
GENCAT Input and Output for Chapter 4   273

REFERENCES   278

INDEX   291

List of Figures

5.1 Observed and Expected (Under Model $\mathbf{X}_5 \boldsymbol{\beta}_5$) Proportions Receiving a Prison Sentence by Type of Offense, Defendant's Income, and Prior Arrest Record   60

11.1 Hypothetical Membership of Five Members Enrolling in Two Successive Years Followed Over a Six-Year Study Period   193

A.1 Graph of $\ln(x)$ 233

A.2 Graph of e^x 234

B.1 Interaction   258

B.2 Observed Sex by Faculty Rank Interaction   264

D.1 Three Methods for WLS Analysis Using the GENCAT 1.2 Program   270

List of Tables

3.1 An $s \times r$ Contingency Table   20

3.2 Table of Probabilities   20

3.3 Status of HMOs Given Grant Funds Before and After PL93-222   21

3.4 Results of a Breathing Test by Smoking and Age Categories   21

3.5 Attitudes Toward School Desegregation by Race and Political Party Identification   28

3.6 Attitudes Toward School Desegregation by Political Party Identification   31

3.7 Rearrangement of Table 3.6   32

4.1 Disposition of Criminal Cases by Prior Arrest Record and Type of Offense   42

4.2 Observed Cell Proportions Based on Frequency Data from Table 4.1   44

4.3 Observed and Predicted Proportions and Residuals[a]   47

4.4 Responses of 201 FDP Members from Berne Canton, Switzerland   52

5.1 Disposition of Criminal Cases by Prior Arrest Record, Type of Offense, and Defendant's Income   56

5.2 Summary of Estimated Parameters and Test Statistics for Model $\mathbf{X}_2\boldsymbol{\beta}_2$ 59

5.3   Definition of Parameters for Model $\mathbf{X}_3\boldsymbol{\beta}_3$ 61

5.4   Summary of Estimated Parameters and Test Statistics for Model $\mathbf{X}_3\boldsymbol{\beta}_3$ 62

5.5   Summary of Estimated Parameters and Test Statistics for Model $\mathbf{X}_4\boldsymbol{\beta}_4$ 64

5.6   Summary of Estimated Parameters and Test Statistics for Final Modular Model $\mathbf{X}_5\boldsymbol{\beta}_5$ 65

5.7   Comparison of Main-Effects Model $\mathbf{X}_1\boldsymbol{\beta}_1$ with Final Modular Model $\mathbf{X}_5\boldsymbol{\beta}_5$ 66

5.8   Response to Mail Questionnaire by Year of Graduation and Physician Type   68

5.9   Byssinotic Response by Condition of Workplace, Length of Employment, and Smoking Habits   70

5.10   Number of Incidents Reported in National Crime Survey   72

6.1   Number of Outpatient Contacts with Pediatric Physicians Made by Families During 1970   77

6.2   Data from Table 6.1 with Subpopulations Having Few Observations Either Combined or Eliminated   78

6.3   Observed Probabilities of Contacts by Age, Sex, and Number of Family Members Enrolled in Plan   79

6.4   Observed, Predicted, and Residual Means for Data in Table 6.2   80

6.5   Number of Pediatric Physician Contacts by Subscriber's Age and Sex, Number of Children Enrolled, and Age of Youngest Child   84

6.6   Observed, Predicted, and Residual Estimates Based on Model $\mathbf{X}_3\boldsymbol{\beta}_3$ 86

6.7   Test Statistics for Model $\mathbf{X}_3\boldsymbol{\beta}_3$ 88

6.8   Number of Pediatric Physician Contacts by Subscriber's Age, Number of Children Enrolled, and Age of Youngest Child   88

6.9   Observed, Predicted, and Residuals Based on Model $\mathbf{X}_4\boldsymbol{\beta}_4$ 90

6.10   Test Statistics for Selected Hypotheses Under Model $\mathbf{X}_4\boldsymbol{\beta}_4$ 90

6.11   Intergenerational Status Mobility Data for British and Danish Father–Son Pairs   93

7.1   PFT Results for People Under Age 40 by Smoking Status and Lead Level in Ambient Air   101

7.2   Probabilities for a Two-Factor, One-Response Study   102

7.3   Parameter Estimates and Test Statistics for Lead and Smoking Status Effects   108

7.4   Prosecutor's Decision to Prosecute Misdemeanors in North Carolina by Offense Charged at Arrest, County of Arrest, and Defendant's Race   112

7.5   Preference of Soldiers for Camp Location by Race, Region of Origin, and Location of Present Camp   114

7.6   Responses of 266 People at Two Points in Time to Questions Pertaining to Voting Intention and Perception of Republican Presidential Candidate   116

8.1   Distribution of Responses to Three Trust Questions by Race and Income   123

8.2   Examples of Table Structures for Multifactor, Single-Response, and Multiresponse Designs   125

8.3   Observed and Predicted Mean Trust Scores   127

8.4   Estimated Parameters and Selected Tests of Significance Under Model $X_1 \beta_1$ 129

8.5   Observed and Predicted Proportions Disagreeing and Agreeing to Three Trust Questions by Race and Income   132

8.6   Selected Test Statistics (X^2) for Hypotheses Under Model $X_4 \beta_4$ 134

8.7   Selected Test Statistics for Hypotheses Under Model $X_5 \beta_5$ 136

8.8   Responses of 693 People to Two Administrations of a Question by Education   139

9.1   Responses to the Statement "The Government Should Make Sure That White and Negro Children Go to the Same School Together" ·   148

9.2   Presence or Absence of Opinion by Education   149

9.3   Computing Gamma Coefficients for Data Reported in Table 9.1 (Panel *a*)   152

9.4   Gamma Coefficients for Data in Table 9.1   156

9.5   Dumping Syndrome by Operation Type and Hospital   160

10.1   Untransformed and Transformed Ranks Assigned to Hypothetical Data   167

10.2   Observed Transformed Mean Scores $\bar{r}^*$ and Estimated Variance-Covariance Matrix V_F for Social Services Data   174

10.3   Summary of Selected Tests of Significance for Mean Spending Scores   176

10.4   Estimates of Spending Preferences for Social Services Programs Based on Reduced Model $X_4 \beta_4$ 178

10.5	Observed Transformed Mean Scores $\bar{r}*$ and Estimated Variance-Covariance Matrices V_F for Social Services by Race	180
10.6	Preference Data for Tax Policy Alternatives by Sex and Ideology	182
10.7	Ordering of Four Questions Pertaining to Postmaterialism and Materialism Values by Age and Education	185
11.1	Retention Experience of 2115 Subscribers Who Joined During 1967–1970 by 6-Month Intervals	195
11.2	Summary of Model Fitting for the 12 Populations	201
11.3	Five-Year Survival Data for 126 Cases with Localized Kidney Cancer	206
11.4	Contraceptive Experiences of 495 White Couples	207
11.5	Survival Rates for Recurrence of Breast Cancer by Three Characteristics of the Cancer	208
A.1	Relationship of x to e^x for Selected Values	233
A.2	Relationship of x to Natural Logarithm of x for Selected Values	233
B.1	Monthly Salary, Sex, and Faculty Rank Data for 18 Faculty Members in the Same Department	239
D.1	GENCAT Computer Setups	271

Public
Program Analysis
A NEW CATEGORICAL DATA APPROACH

PART I

Foundations

1

Multidimensional Contingency Tables

This book introduces workers in public health and public affairs to a general analytic methodology proposed by Grizzle, Starmer, and Koch (1969). This methodology is based on applications of the *general linear model* (the basis for both regression analysis and analysis of variance for continuous data) to categorical data. By *categorical data* we mean information measured on nominal or ordinal scales or grouped continuous data. Appendix B presents the general linear model in the analysis of variance situation; this appendix serves as a guide for the use of the general linear model with categorical data.

Researchers in public health and public affairs often encounter categorical data in the analysis of program performance. It occurs in many places—administrative files, voting records, public opinion surveys, treatments in clinical trials, crime reports, and census data, to name a few. Health planners, for example, study the relationships between social characteristics and the demand for services. Crime analysts seek to establish the characteristics of personal and property crimes and to determine how the criminal justice system disposes of felony arrests. Opinion researchers try to detect trends in attitudes.

Until recently there were no general approaches to categorical data. The emphasis was on the analysis of the two-way table, and nothing beyond this was taught in introductory courses. Only a few books dealt with categorical data analysis (Maxwell, 1961; Cox, 1970; Fleiss, 1973), and none focused primarily on the analysis of data from public programs. With the evolution of the high-speed computer, three general approaches to categorical data appeared: the *maximum likelihood* (ML) approach, the *minimum discrimination*

3

information theory, and the *weighted least squares* (WLS) approach. All are equivalent asymptotically—that is to say, when the sample size is infinite—and they also yield similar results when the sample size is large. Among the recent books that focus on the maximum likelihood method are Bishop, Fienberg, and Holland (1975), Fienberg (1981, second edition), Everitt (1977), Haberman (1978), and Goodman and Magidson (1978), while Gokhale and Kullback (1978) have described the information theory approach. This book presents the weighted least squares approach developed by Grizzle, Starmer, and Koch (1969). This methodology, besides providing an alternative to the other approaches, can be used in situations where the maximum likelihood method and information theory require the solution of complex mathematical expressions.

In this book we demonstrate the flexibility of the WLS approach. We do not present the entire range of its applications, however, for there are too many different applications for a single book and they sometimes present complexities beyond the scope of an introductory work. Because the use of WLS in categorical data analysis is a recent development, the subject is still expanding and new contributions are continually appearing. In the final chapter, Chapter 12, we briefly discuss some additional applications and provide references to the pertinent WLS literature.

The WLS approach is ideally suited to the analysis of *multidimensional contingency tables*—tables that are formed by the crosstabulation of a number of categorical variables. In the analysis of multidimensional contingency tables, a large number of functions of the variables may be of interest. Examples of the functions analyzed in this book include the proportion of defendants in a criminal trial receiving a prison sentence (Chapters 4 and 5), the mean number of visits to a health maintenance organization (Chapter 6), the rank correlation between responses to two attitudinal questions (Chapter 9), and the adjusted probability of membership retention in a health maintenance organization (Chapter 11). The table inside the front cover describes the functions analyzed in each chapter. Once we have selected the functions for analysis, the next step is to determine how these functions vary across the levels of the other variables—for example, how does the proportion of defendants receiving a prison sentence vary by race and type of crime? In theory we can examine the variation of the selected functions across an unlimited number of combinations of levels of the other variables. In practice the number of combinations of levels of the other variables is limited by computer size.

This book has three parts. Part I presents the WLS approach developed by Grizzle, Starmer, and Koch as well as the guidelines and philosophy used in the analyses in Parts II and III. Part II contains relatively simple applications of the WLS methodology; each chapter presents a slightly more involved application than the previous chapter. Part III provides more complex applications that demonstrate the range and flexibility of the WLS procedure. Now let us take a closer look at the book and its contents, chapter by chapter.

Chapter 2 discusses our philosophy of data analysis and presents some guidelines for the subsequent analyses. There is also a discussion of the choice of scales of measurement to be used in the analysis.

Chapter 3 introduces the WLS approach developed by Grizzle, Starmer, and Koch. We explain the mathematical ideas by analogy to situations familiar to the reader and offer several simple applications of WLS methodology.

Chapter 4 provides the first detailed application of WLS approach to data. We use data from a North Carolina criminal court to examine a simple linear function—the probability of conviction—and its relationship to the type of offense charged and the defendant's prior arrest record. Our emphasis is on fitting the linear model and interpreting the model parameters.

In Chapter 5 we use the same substantive problem introduced in Chapter 4 but include an additional variable, the defendant's income level, in order to discuss a more complex model and the treatment of additive interaction. Because the interaction effects do not allow for a parsimonious presentation, a "modular" model is developed as an alternative. The emphasis in this chapter is on testing for the presence of interaction and developing the modular model.

In Chapter 6 we move from dependent variables having two categories to variables with more than two categories. In the health planning example we consider, the levels of the response variable are ordered. This ordering allows us to construct a mean score for the various population subgroups, and these mean scores are then analyzed by using the linear model framework.

Chapter 7 introduces log-linear models. The logit function is used in the analysis of the relationship of a lung function test to the subject's smoking habits and the level of environmental pollution. The use of the logit function allows us to compare the weighted least squares and maximum likelihood estimates.

Chapter 8 introduces a multiresponse analysis, that is, the analysis of more than one function within a subpopulation. The level of trust toward three government institutions—the President, the Supreme Court, and the United States Senate—expressed in a national opinion survey is analyzed, and the differences in trust are examined in relation to the respondents' income and race.

Chapter 9 gives the first example of the use of a complex function: Goodman and Kruskal's rank correlation coefficient gamma. We use survey data from the Houston Community Study to analyze the effects of question wording and education on consistency between two responses.

In Chapter 10 we use data from the Houston Community Study to present another multiresponse—that is, multiple dependent variables—analysis. The preferences of a sample of Houstonians toward various policy alternatives are measured by having respondents rank alternatives by allocating a finite amount of dollars to a fixed set of alternatives. We then fit models that reveal whether citizens are indifferent (have no clear preferences) toward various subsets of alternatives.

In Chapter 11 we use WLS methodology to create statistics from a follow-up life table in order to analyze the retention of members enrolled in a health maintenance organization (HMO). This application of "survival curve" techniques has direct applications for planners interested in service delivery.

Chapter 12 reviews additional WLS applications and discusses recent theoretical developments that extend the range of problems for treatment within the WLS framework.

The WLS methodology uses matrices combined with the linear model approach. To aid the reader who is unfamiliar with either of these fields we have provided two appendixes. In Appendix A we define matrix terminology, matrix arithmetic, and the use of matrices in solving a system of linear equations. The exponential and natural logarithmic functions are also reviewed. Appendix B introduces the general linear model and its use in the analysis of variance. Special emphasis is given to model parameterization and statistical interaction. It is essential that the reader be familiar with matrix multiplication and the design matrix concept reviewed in these appendixes before reading Chapter 3.

There are two other appendixes: Appendix C contains a table of chi-square values. Appendix D describes the use of the GENCAT computer program to perform weighted least squares analyses. Listings of the GENCAT computer setups used to perform the analyses in Chapters 3 through 11 are found at the end of those chapters.

In summary, problems encountered in public health and public affairs often involve more than two categorical variables, and thus, there is a need for a methodology designed for the analysis of multidimensional contingency tables. The flexible and powerful WLS approach developed by Grizzle, Starmer, and Koch is one such method. In this book we demonstrate how this approach is applied to real (and often troublesome) data from the fields of health and public affairs. Because of our practical experience, we stress applications but do not ignore the theory. Therefore, before describing the WLS method and explaining how it is applied, we turn to the philosophy and ground rules that guide the analyses.

2
Guidelines for Analysis

In this chapter we describe the situations in which the WLS approach for categorical data can be used. We also define the type of functions involved and offer some guidance on their use. Lastly, we discuss different approaches to data analysis and model building and make some general recommendations.

Though statistics is taught as a science, there is art in its practice—and hence differences of opinion. Not everyone will agree with our guidelines and recommendations, but at least they give the reader a point of departure. We begin this chapter by defining categorical data.

WHAT ARE CATEGORICAL DATA?

Statistical terminology is not standardized and, as a result, various writers use different words to refer to the same type of data. The following table summarizes the terms used by some authors and disciplines to define "categorical" data:

Level of Measurement	*Sociology/Political Science*	*Fienberg* (1977)
Nominal	Qualitative or nonmetric; attribute data	Dichotomous; nonordered polytomous
Ordinal	Qualitative or nonmetric; attribute data	Ordered polytomous
Grouped interval	Quantitative or metric	

Regardless of the terms used to describe categorical data, the information is measured such that it is classified into discrete categories. These categories may be ordered (ordinal or grouped interval) or unordered (nominal). In the case of grouped interval data, the categorical variable is derived from a continuous scale and it is therefore possible to determine the distance between the midpoints of the categories.

One may ask why a continuous variable would be converted into a grouped interval, or even ordinal, variable with the resulting loss of information. The answer depends on the concept being measured and the process used to record the information. People in the social and health sciences often find that the method used to operationalize a concept is not directly related to the construct being measured. Take, for example, the variable "years of formal education completed." This variable results in the assignment of the positive integers 0, 1, 2, . . . to the observational units. Many studies would treat this variable as continuous, thereby implying that the distances between each level (1 year) are equal. For some applications—estimating the number of formal years of schooling per capita, for example—this approach is perfectly acceptable. But many sociological concepts, such as educational attainment, are not so tractable as years of formal schooling. Sociologists observe that the *social distance* between years of schooling is not captured by this scale. The distance between completing high school (12 years) and the first year of college (13 years) is substantially different from the 11–12-year interval for most indicators of human behavior. Educational attainment represented by years of schooling often is mapped into a categorical variable with the following categories:

- Grade school or less (0–8 years)
- Some high school (9–11 years)
- Completed high school (12 years)
- Some college (13+ years)

Although analysts differ on the definition of the categories, they usually reduce the information presented by years of schooling to an ordinal scale. Most would agree that the midpoints of this scale do not represent meaningful distances of educational attainment.

This example is representative of many choices made about measurement and analysis in the social and health sciences. Such choices are based on the social theory underpinning the analysis and are not simply arbitrary preferences. The advantage provided by a generalized categorical methodology is that comprehensive multivariate analyses are still possible even though the level of measurement of some variables has been *reduced*—that is, defined on a categorical scale instead of a continuous one. In the past, analysts were often faced with the choice either of creating a fiction regarding the level of measurement in order to use continuous data multivariate techniques or of using bivariate nonparametric methods that could not reflect the multivariate structure of the problem (see Siegel, 1956). The development of multivariate categorical methods, of which the WLS approach is an example, provides the analyst with alternatives. Now that we know what categorical data are, the next section discusses the problems we can analyze with categorical data techniques.

IS THIS A CATEGORICAL DATA PROBLEM?

The number and type of categorical variables determine the choice of analyses available to the researcher. We characterize variables as *dependent* (also called response) or *independent* (also called factor, explanatory, or predictor) variables according to the role they play in the analysis. Problems may also be characterized by their number and type of measurement (all categorical, all continuous, both categorical and continuous). These two dimensions lead to the following analysis situations:

		INDEPENDENT VARIABLES		
		All Categorical	*All Continuous*	*Both Categorical and Continuous*
Dependent Variables	*All Categorical*	(*a*) Categorical data analysis	(*b*) Logistic regression	(*c*) Logistic covariance analysis
	All Continuous	(*d*) Analysis of variance (dummy variable regression)	(*e*) Regression	(*f*) Analysis of covariance

The WLS approach is most useful in the pure categorical situation (case *a*). It may also be applied to other situations (cases *b–c*) classified in the first row of this scheme provided that one groups the continuous variables. At present no ideal general method for analyzing problems classified in categories *b* and *c* exists, although some alternatives have been proposed by Walker and Duncan (1967), Cox (1970, 1972a), and Nelder and Wedderburn (1972). The remaining three cells (*d–f*) may be handled by using traditional parametric techniques with which the WLS approach enjoys a close mathematical relationship. Once we know that the problem is appropriate for categorical data analysis, we have to decide what *type* of analysis should be performed.

WHAT FUNCTION SHOULD BE ANALYZED?

One of the advantages of the WLS approach is the discretion that the technique provides the analyst in choosing a function of the dependent variable for analysis. This function may be a simple proportion, a mean score, or a more complex measure (such as a rank correlation coefficient). *The WLS methodology permits the function to be formed by a linear, logarithmic, or exponential transformation or any combination of these three operations.* Although the methodology permits considerable flexibility in the choice of function, interesting and useful analyses can be performed using very simple functions. We have relied primarily on simple response functions throughout most of the book.

Chapters 4, 5, 6, 8, and 10 are based on simple *linear* transformations of the categorical information. Chapters 4, 5, and 8 use proportions to characterize the dependent variable. Chapters 6, 8, and 10 contain analyses based on linearly weighted sums, or mean scores, of the probabilities.

Apart from these applications of linear functions, there may be times when we prefer to use either the *logarithmic* scale or the *logit* (logistic) scale. If we believe that the probability of an event is generated by the product of two independent factors—that is,

$$p_{ij} = c \times a_i \times b_j \tag{2.1}$$

where c is a constant and a_i and b_j are the effects of the factors, then the *logarithmic* scale has an appeal. Taking the natural logarithm of both sides of equation (2.1) yields.

$$\ln(p_{ij}) = \ln(c) + \ln(a_i) + \ln(b_j) \tag{2.2}$$

This transformation has converted the multiplicative or product model to an additive one in terms of the logarithms of the factor effects, and hence we can use the methodology presented in Chapter 3 to analyze the data on the logarithmic scale.

It appears that the logarithmic transformation is especially useful when the response variable has only two levels, say disease or no disease. Since these categories are mutually exclusive and exhaustive, their probabilities sum to 1, and therefore knowledge of one of these probabilities implies knowledge of the other. Although one should be able to focus attention on the logarithm of either probability and use it in estimating the logarithm of the effects of the factors, this in fact is not the case. An analysis based on log "disease" will not necessarily provide the same conclusions as an analysis based on log "no disease." This difference in the results poses no problem when there is a clear choice as to which probability should be used, but an obvious choice is often the exception rather than the rule. Thus we will not be using the simple logarithmic transformation in this book.

Chapter 7 introduces another form of log-linear modeling using the *logit* function. Although the logit is a simple function, it requires both linear and logarithmic transformations of the frequencies of the dependent variable. Results based on logit analysis are independent of response level (disease or no disease) selected. Moreover, one can use the logit scale to generate the hypothesis of multiplicative association described in Chapter 3.

The logit transformation of the probability p is

$$\text{logit}(p) = \ln\left(\frac{p}{1 - p}\right) \tag{2.3}$$

At first glance this looks like a strange transformation to make. As will be shown later, however, the logit transformation of the probabilities provides a reasonable approach to the analysis of response variables that have two levels. Logit models have the technical advantage of always yielding estimated proportions in the range 0.0 to 1.00, whereas models using a sample proportion may produce estimated proportions beyond this range—particularly when the analysis is based on small samples. This problem of estimation beyond the range has no practical consequence whenever the analysis is based on large samples, since the goodness-of-fit statistic for the model is usually significantly large to suggest consideration of alternative models. (See Koch, Gillings, and Stokes, 1980, pp. 202–203.)

Results of the analysis on the logit scale may differ from results based on the linear scale. This difference is not unexpected, since they are based on different functions of the probabilities. In some cases the logit transformation may provide a simpler model—that is, a model with fewer interaction terms—while in other cases the linear model may yield the simpler model. The choice of whether to use the linear or logit scale is left to the investigator.

Often the response level has more than two levels and we wish to create a *complex* function. In these cases, the WLS methodology is particularly appropriate because of its flexibility in constructing functions, whereas techniques involving maximum likelihood estimation are not directly applicable.

Chapters 9 and 11 introduce two examples of complex functions: a rank correlation coefficient (gamma) and an adjusted probability of survival function. These functions require the use of a combination of linear, logarithmic, and exponential transformations. They are the most complex functions introduced in this book, but they only suggest the range of functions that the analyst may form. Chapter 12 and the References point to other analyses using complex functions.

The choice of function has two important implications: *One is for the procedure used to estimate model parameters; the other is for the interpretation of model results.* The choice of proportions, mean scores, rank correlation coefficients, or more complex functions limits one's analysis to the use of WLS estimation, since programs for the maximum likelihood approach to estimation are not readily available for the functions described in this book (apart from the logit or logarithmic transformations). Two computer programs—GENCAT, a computer program developed by Landis and others (1976), and FUNCAT

(Barr, Goodnight, and Sall, 1979), a program in the SAS program package—may be used to produce the WLS estimates.

The choice of function also affects one's *interpretation* of model results. Log-linear models permit examination of the model effects based on the *ratios* of the cell frequencies, whereas linear models make comparisons based on their *differences*. Interpretation of the model parameters differs in each case.

Two articles by Bhapkar and Koch (1968a, 1968b) provide a general overview of the types of hypotheses that one may choose to test when working with categorical data. They provide examples of analyses on both the additive and multiplicative scales, and they also make the point that the investigator can choose the scale that seems most appropriate for the data.

Before performing any analysis we must ensure that our experiment will yield enough data for analysis. In the following section, therefore, we discuss the issue of sample size.

WHAT SAMPLE SIZE IS REQUIRED?

The WLS method, because of the asymptotic nature of its test statistics, imposes requirements regarding minimum sample size. The various statistics produced by the WLS estimation procedure will be well behaved whenever they are based on a sufficient sample size. When sample size requirements are met, the test statistic is distributed as central chi-square if the hypothesis being tested is true. When the sample sizes are small, the analyst cannot be certain that the *p* values associated with the test statistics represent the true probability of falsely rejecting the null hypothesis.

Because the WLS method applies to a broad range of situations, we have not attempted to specify sample size guidelines for all conditions. Rather, we have focused on the more common (and less complicated) situations presented in this book. In the paragraphs below we specify guidelines for the following situations:

1. Analyses of means and proportions or logits where the probability of success (π) is common $(0.2 \leq \pi \leq 0.8)$

2. Analyses of proportions or logits where the probability of success (π) is extreme $(\pi < 0.2$ or $\pi > 0.8)$

3. Multiresponse situations

   The guidelines that follow are not absolute; they represent conditions that yield sound statistical conclusions in most cases. Violating one or more of the guidelines simply introduces a greater degree of uncertainty about the soundness of the conclusions but does not mean necessarily that the analysis is invalid.

Means and Proportions or Logits Based on Commonly Occurring Events

When a single function such as the proportion or logit is constructed for each subpopulation, the following guidelines for sample size apply:

1. Ideally, no more than one-quarter of the functions should be based on subpopulation samples of less than 25 cases.

2. No subpopulation sample size should fall below 10 cases.

These guidelines apply to the analyses presented in Chapters 4, 5, 6, and 7. The following example—showing acceptable sample size—illustrates an application of these rules:

Sex and Region	Yes	No	Total
Male north	8	18	26
Male south	3	7	10
Female north	10	15	25
Female south	14	16	30

This example meets the minimum sample size requirements for WLS analysis using a mean, proportion, or logit. Three subpopulations have at least 25 cases, the remainder has at least 10 cases, and the probabilities of success are not extreme.

Proportions or Logits Where the Probability of Success (π) Is Uncommon ($\pi < 0.2$ or $\pi > 0.8$)

Extreme events require more observations per subpopulation to ensure adequate estimation. Given the subpopulation parameter π_i (the probability of success for subpopulation i) and sample size n_i, the following rule should apply:

$$n_i \pi_i \geq 5 \quad \text{and} \quad n_i(1 - \pi_i) \geq 5$$

This guideline is conservative; some authors (see Cochran, 1954, and Lewontin and Felsenstein, 1965) propose less stringent requirements.

Multiresponse Situations

When one or more functions are constructed *within* each subpopulation, an additional requirement applies. Because the *same* observations are used to estimate *multiple* functions, a larger sample size is necessary. If m functions are created per subpopulation, the required subpopulation sample size determined above is multiplied by $\sqrt{m}$. As an example, Chapter 8 presents an analysis where six simple proportion functions are estimated for each of five subpopulations. According to our guidelines, we should have at least 25 cases for most of the subpopulations. Since six functions are estimated in each subpopulation, most of the subpopulations should have a minimum of $\sqrt{6} \times 25$ or about 62 observations in each subpopulation. If the probability of the event had been extreme (say $\pi = 0.10$) and six proportion functions were estimated for each subpopulation, recommended sample size would be at least $50 \times \sqrt{6}$ or about 123 cases per subpopulation. Chapters 8 and 10 present situations where the multiresponse rules apply.

Further discussion of sample size requirements is provided by Landis, Heyman, and Koch (1978) and by Koch and others (1977). So far we have discussed totals for the subpopulations; in the following section we address specific cells.

ARE ZERO CELLS A PROBLEM?

There are two situations in which a zero frequency count causes estimation problems. The first occurs when the function created is a proportion and the observed value of the proportion is either 0.0 or 1.0. In this case the estimated variance of the observed function is zero. Since the variance appears in the denominator of the chi-square test statistic and since division by zero is not defined, the test statistic is undefined. We should also note a philosophical problem associated with an estimated variance of zero: It implies certainty about the value of the function, but certainty never exists when the observed value is based on a sample of observations.

The second situation where a zero frequency count poses problems occurs whenever a logarithmic or logit transformation is performed on an observed value of zero. Since the logarithm of zero is undefined, one cannot compute the function and its observed variance. This problem arises in Chapter 10 during the computation of Goodman and Kruskal's gamma function.

Grizzle, Starmer, and Koch (1969) recommend the *1/r rule* as a possible solution for both zero cell situations. The $1/r$ rule requires the replacement of the zero frequency cell in the contingency table by the estimate $1/r$, where r is the number of response categories. Not all zero cells in a table require replacement with the value $1/r$—only those cells that present a specific problem. In Chapter 10, for example, only a small number of the zero cell values are changed to permit computation of Goodman and Kruskal's gamma.

Replacement of the zero frequency with a small, nonzero value introduces a slight but acceptable bias into the estimate. The degree of bias depends on the total number of cases in the table, the nature of the function analyzed, and the type of analysis. The addition of a small constant to cells in a contingency table is not unusual. For example, Goodman (1970) suggests that 0.5 be added to every cell in the table regardless of the number of observations in a cell. Bhapkar (1979) recommends the substitution of a number of the size $1/n_i$ for zero cells in the analysis of linear functions and the addition of a constant to all cells for logarithmic functions.

An alternative to using the $1/r$ rule is to restructure the problem by combining or eliminating columns (response categories) and rows (subpopulations). This can be done if the recombination of the rows or columns makes sense substantively. Moreover, the rows or columns that are combined should have similar distributions.

The analyst who works with self-weighting or proportional sampling and relies on secondary analyses of archival data often finds that the sample size is too small or the data are grouped in one or two cells of each row. This situation is common in the analysis of social survey data—for example, a proportional national sample of the voting preferences of Americans classified by party identification, race, and social status will produce very few southern black Republicans. We recommend that the analyst either pool the subpopulations having insufficient sample size with a conceptually similar group or eliminate the subpopulation entirely from the analysis. The latter approach is applied in Chapter 8. When we eliminate selected groups for whom essentially no information is available and thus nothing is known, the analysis may proceed without a substantial loss of generality. The major cost associated with aggregation or elimination of cases is to restrict our ability to estimate some high-order interaction terms and test certain hypotheses.

Now that we have the proper sample size and know what type of analysis to perform, we still must determine how many analyses we are going to perform.

HOW MANY ANALYSES SHOULD BE PERFORMED?

Most data analysis can be placed on a continuum. At one end is a strict statistical position regarding the use of tests of significance; at the other is an exploratory strategy in which tests of significance are but one tool leading an analyst to decisions regarding the data. The strict position limits tests of significance to propositions formulated prior to testing and, ideally, prior to data collection. The exploratory strategy often formulates hypotheses during analysis and on the basis of previous exposure to the data. Often the data analyzed were not collected for the present purpose, and the analyst, in retrospect, must face questions about sampling and hypothesis formulation.

Each approach has its advantage and limitations. The strict testing approach strengthens statistical rigor and protects the analyst against introducing personal bias into the analysis. It also severely limits the potential payoffs by confining the analysis to anticipated questions; thus it limits serendipity. It also requires considerable knowledge about a subject to formulate suitable hypotheses. The exploratory approach requires little prior knowledge about the data and often produces unexpected results, but it leaves the analyst open to the mistake of affirming the results as important findings when they may be in fact simple artifacts of the data set under study. Most analyses fall in the middle of this continuum rather than at the end points; rarely does one do only strict testing or only an exploratory analysis.

We encourage the analyst to undertake strict tests whenever the state of knowledge and data permit. When this is not possible, we advocate "listening to the data" within limits. Conducting every possible test is neither productive nor desirable. The best strategy is to fit a limited number of models that have compelling substantive justification and are readily interpretable.

The exploratory approach is most suitable after initial model testing produces basic findings or suggested unanticipated ones. The exact nature of the "final" model may be suggested by a look at the data, but the ultimate test is its plausibility and interpretability. Since many areas of study are not sufficiently advanced to permit analyses involving only strict testing, we have demonstrated the exploratory approach to data analysis in several chapters.

The exploratory approach to testing has a major drawback regarding determination of the true p value of a particular test. In our attempts to uncover the relationships among variables, we examine a limited number of different models. If the first model does not fit the data, we try another model that incorporates results from our first attempt and continue in this fashion. This procedure represents a stagewise approach to model building; each stage is conditional on what was learned previously. The tests of significance are therefore ex post facto (after stage 1)—which means that the significance levels are actually larger than set originally. Because the true significance level for these ex post facto tests is unknown, the test statistics are used as guides. Knowledge of the substantive problem and the data set at hand guide the interpretation of what is plausible. Because of the problem of determining the true p value of a test, for some analyses we report test statistics without estimates of their associated p values.

Although the levels of significance in some situations are used only as a guide, what levels should be used? In the next section we address this question.

WHAT LEVELS OF SIGNIFICANCE SHOULD BE USED?

In the analysis of data we first attempt to fit a model that characterizes the data adequately. If the model provides a good fit to the observed data, our next step is to test hypotheses about the terms (parameters) in the model. In Chapter 3 we present a statistic for testing the fit of the model as well as a statistic for testing whether a term should be included in the model. Note that we approach the tests of these two types of hypotheses with different goals in mind. In the test that must be performed *first*—the test of the fit of the model—we want to be sure that we have an adequate portrayal of the data before proceeding with the tests about the terms in the model. If there is not an adequate fit to the data, we would not usually go forward with the other tests because there is too much unexplained variation to have confidence in their results.

Since we wish to be confident that the model fits, we conduct our test of the model's goodness of fit at a significance level that allows us to detect even slight departures. We believe that using $\alpha = 0.25$—that is, having a 25 percent chance of incorrectly rejecting the true model allows even slight, but real, departures to be detected.

When we have a model that fits, we want it to exhibit parsimony—that is, we want to include only those terms which provide strong evidence that they are real. We therefore recommend the use of $\alpha = 0.05$ if we are testing a single hypothesis and $\alpha = 0.01$ if we are testing several hypotheses based on the same data. We could use more elaborate multiple comparison procedures when testing several hypotheses; however, we feel that in most situations one analyzes the data in stepwise fashion, and the tests of significance are used as a guideline rather than as formal tests of significance. The use of alpha $= 0.01$ is a convenient substitute for formal approaches to multiple comparisons testing. For those interested in formal approaches to multiple comparison procedures, see appendix 6.c in *Statistics for Experimenters* by Box, Hunter, and Hunter (1978).

SUMMARY

In this chapter we defined categorical data and indicated the types of variables and sample size that can be dealt with by the WLS approach. We also made some recommendations regarding the choice of function, the number and strategies of analysis, and the choice of significance levels for the goodness-of-fit test and other tests.

The key topics discussed in the chapter are:

- *Categorical data:* Nominal, ordinal, or group interval.
- *Variable types:* Dependent and independent; categorical and continuous.

  Functions: Linear—easy to interpret, estimates possibly out of range; *logarithmic*—not unique; *logit*—harder to interpret but no problem with estimates.
- *Sample size:* Subpopulation totals ideally greater than or equal to 25 and even larger for extreme values of the probabilities or for multiple functions.

- *Zero cell:* Can be a problem if a logarithmic transformation is performed on it or it yields a function with a zero variance; add a constant.
- *Analysis strategy:* Have a limited number of possible models in mind.
- *Levels of significance:* Goodness of fit—use $\alpha = 0.25$; tests about terms under the model—use $\alpha = 0.05$ or 0.01.

Now that we have presented the guidelines for analysis, we are ready to discuss the WLS theory and its applications.

3

Contingency Table Analysis: The WLS Approach

This chapter introduces the weighted least squares (WLS) method of categorical data analysis, also known as the GSK approach, developed by Grizzle, Starmer, and Koch. We begin with a definition and several examples of contingency tables and then discuss different functions of the contingency table frequencies that we may form. Next we present the variance-covariance matrix of a function and then show its use in applying the linear model to categorical data. Note that the development of the linear model to categorical data follows closely the presentation in Appendix B. The next sections define additive and multiplicative association; this discussion is followed by a demonstration that the hypothesis of no multiplicative association is equivalent to the hypothesis of independence of rows and columns in two-way tables. We begin with a discussion of the contingency table.

THE CONTINGENCY TABLE

A *contingency table* is the result of the cross-classification of two or more nominal, ordinal, or grouped continuous variables with one another. Table 3.1 shows a contingency table with s rows and r columns. The rows, called *subpopulations*, represent combinations of independent or factor variables from which independent random samples of a priori fixed sizes $n_{1.}$, $n_{2.}, \ldots, n_{s.}$ ($n_{i.}$ is the total of the ith subpopulation) have been selected. The columns represent the levels of the dependent or response variables, and the subjects in each subpopulation are classified into one of these response levels. The

19

Table 3.1   An $s \times r$ Contingency Table

	RESPONSE CATEGORIES				
Subpopulations	*1*	*2*	$\cdots$	*r*	*Total*
1	n_{11}	n_{12}	$\cdots$	n_{1r}	$n_{1\cdot}$
2	n_{21}	n_{22}	$\cdots$	n_{2r}	$n_{2\cdot}$
$\vdots$	$\vdots$	$\vdots$	$\vdots$	$\vdots$	$\vdots$
s	n_{s1}	n_{s2}	$\cdots$	n_{sr}	$n_{s\cdot}$

n_{ij} are the number of subjects in the ith subpopulation with response level j as an attribute.

Corresponding to the frequency table in Table 3.1 is the table of probabilities shown in Table 3.2, where π_{ij} is the probability that a subject in the ith subpopulation has the jth response attribute. Note that the probabilities in each row sum to 1. This outcome results from viewing each level of the factor combinations as a distinct subpopulation (that is, the $n_{i\cdot}$ have been fixed in advance) and calculating the probabilities within each subpopulation. An estimate of π_{ij} is p_{ij}, where $p_{ij} = n_{ij}/n_{i\cdot}$.

Table 3.2   Table of Probabilities

	RESPONSE CATEGORIES				
Subpopulations	*1*	*2*	$\cdots$	*r*	*Total*
1	π_{11}	π_{12}	$\cdots$	π_{1r}	1
2	π_{21}	π_{22}	$\cdots$	π_{2r}	1
$\vdots$	$\vdots$	$\vdots$	$\vdots$	$\vdots$	
s	π_{s1}	π_{s2}	$\cdots$	π_{sr}	1

Table 3.3 provides an example of 2 × 2 table. The row variable indicates periods before and after the passage of PL93-222 (the Health Maintenance Organization Act of 1973); the column variable indicates the number of HMOs that had or had not terminated activity by 1976 (Strumpf and Garramone, 1976).

Table 3.4 provides an example of a more complex table: a three-way table with one response and two factor variables. The rows of the table result from the combination of the smoking status and age variables whereas the

Table 3.3   Status of HMOs Given Grant Funds Before and After PL93-222

Time Period	STATUS		Total
	Terminated	*Not Terminated*	*Total*
Before PL93-222	37	42	79
After PL93-222	41	67	108

Table 3.4   Results of a Breathing Test by Smoking and Age Categories

Age	Smoking Status	BREATHING TEST RESULTS			Total
		Normal	*Borderline*	*Abnormal*	*Total*
<40	Never smoked	577	27	7	611
	Former smoker	192	20	3	215
	Current smoker	682	46	11	739
40–59	Never smoked	164	4	0	168
	Former smoker	145	15	7	167
	Current smoker	245	47	27	319

columns represent the results of a breathing test given to Caucasians in nine different industrial plants in Houston, Texas, during 1974 and 1975 (McClimans, 1977).

Table 3.4, as noted, contains one response (breathing test) and two factor variables (age and smoking status), and the combination of age and smoking status yields six subpopulations. The row totals for these six subpopulations are supposed to be fixed a priori, but the row totals in this situation (that is, 611, 215, . . . , 319) do not look like numbers that might be selected in advance. The row totals here are actually the number of subjects of the 2219 total that were observed in each age by smoking status category. We did not select people until we found 611 subjects less than 40 years old who have never smoked; rather, 611 is the number of people we observed in that category.

When it is reasonable to believe that the levels of the response variable may be related to the combinations of the factor variables, however, we may treat the row totals as if they had been fixed in advance without changing the results of the study. Whether we treat these data as one response and two factor variables or as three response variables is unimportant to the results. Here we have chosen to view the table as having one response and two factors because breathing test results may depend on age and smoking status and we want our data layout to reflect that possibility. See Bhapkar and Koch (1968a, 1968b) for a detailed discussion of this topic.

To perform the analyses presented in the following sections, we assume that the observed data in each subpopulation follow the *multinomial* distribution—an extension of the familiar binomial distribution for situations with more than two possible outcomes. This assumption, unlike that of the assumption of the normal distribution for continuous data, presents no limitation on the analysis if the observed data in each subpopulation come from a simple random sample from a large population. In this case, we may then assume the underlying multinomial distribution. The WLS approach may also be used for sampling processes more complex than simple random sampling. The estimation methods applicable for complex samples, which are beyond the scope of this book, are reviewed in Chapter 12.

In the following sections we discuss the analysis of contingency tables, beginning with the formation of the functions of the cell probabilities used in the analysis.

FORMATION OF FUNCTIONS

In the analyses described in Parts II and III, we examine the relationships between functions of the probabilities, **F**, and the independent or factor variables. The functions may be quite simple (for example, the probability itself) or complex (for example, a rank correlation coefficient between two response variables). The choice of the function to be used in the analysis depends on the investigator's interest and goals.

The functions of the probabilities may be used to test hypotheses directly or as an intermediate step in the analyses. In the latter case, the functions formed are used as the dependent variables in a linear model analysis. The goal of the analysis is the discovery of the relationship between the dependent variable and the factor variables. We focus on this use of the functions in Parts II and III. Later in this chapter we also show how the functions can lead to the direct testing of hypotheses.

The functions that can be formed involve linear, logarithmic, exponential, or combined transformations of the cell probabilities. Matrix theory allows these different functional types to be represented succinctly. To see this we let π represent the vector of probabilities from Table 3.2, where

$$\pi = \begin{bmatrix} \pi_1 \\ \pi_2 \\ \vdots \\ \pi_s \end{bmatrix} \quad \text{and} \quad \pi_i = \begin{bmatrix} \pi_{i1} \\ \pi_{i2} \\ \vdots \\ \pi_{ir} \end{bmatrix}$$

A *linear transformation* of the probabilities can be represented by $\mathbf{F} = \mathbf{A}\pi$, where the matrix $\mathbf{A}$ contains the coefficients of the π_{ij} in the linear transformation. For example, we may wish to examine the proportion of HMOs terminated minus the proportion not terminated for the two time periods for the data in Table 3.3. This function may be expressed as

$$f_1 = 1 \cdot \pi_{11} - 1 \cdot \pi_{12} + 0 \cdot \pi_{21} + 0 \cdot \pi_{22} = \pi_{11} - \pi_{12}$$
$$f_2 = 0 \cdot \pi_{11} + 0 \cdot \pi_{12} + 1 \cdot \pi_{21} - 1 \cdot \pi_{22} = \pi_{21} - \pi_{22}$$

or rewritten in matrix terms as

$$\mathbf{F} = \begin{bmatrix} f_1 \\ f_2 \end{bmatrix} = \mathbf{A}\pi = \begin{bmatrix} 1 & -1 & 0 & 0 \\ 0 & 0 & 1 & -1 \end{bmatrix} \begin{bmatrix} \pi_{11} \\ \pi_{12} \\ \pi_{21} \\ \pi_{22} \end{bmatrix}$$

where the elements in the first row of $\mathbf{A}$ are the coefficients of the π_{ij} in f_1 and the elements in the second row of $\mathbf{A}$ are the coefficients of the π_{ij} in f_2. Other linear functions that may be chosen are a cell probability itself (discussed in Chapters 4 and 5) or a weighted sum of probabilities (discussed in Chapter 6).

Logarithmic functions, another class of functions treated by the WLS approach, may be succinctly written in matrix notation as $\mathbf{F} = \mathbf{A}_2 \ln(\mathbf{A}_1\pi)$. Suppose that we wish to form the logit of being terminated for the data in Table 3.3. Recall that in Chapter 2 the logit is defined as

$$\text{logit}(\pi_{ij}) = \ln\left(\frac{\pi_{ij}}{1 - \pi_{ij}}\right)$$

Since $\pi_{i1} + = 1$, we know that $\pi_{i2} = 1 - \pi_{i1}$ and therefore we have

$$\text{logit}(\pi_{i1}) = \ln\left(\frac{\pi_{i1}}{\pi_{i2}}\right)$$
$$= \ln(\pi_{i1}) - \ln(\pi_{i2})$$

Hence we must form a linear combination of the logarithm of the probabilities—that is,

$$f_1 = \ln(\pi_{11}) - \ln(\pi_{12})$$
$$f_2 = \ln(\pi_{21}) - \ln(\pi_{22})$$

and this is expressed in matrix terms as

$$\mathbf{F} = \begin{bmatrix} f_1 \\ f_2 \end{bmatrix} = \mathbf{A}_2 \ln(\mathbf{A}_1 \boldsymbol{\pi})$$

$$= \begin{bmatrix} 1 & -1 & 0 & 0 \\ 0 & 0 & 1 & -1 \end{bmatrix} \ln(\mathbf{A}_1 \boldsymbol{\pi})$$

Since all four of the π_{ij} are used in this function we let $\mathbf{A}_1$ be the identity matrix. Hence the transformation simplifies in this case to

$$\mathbf{F} = (\text{logit}(\pi_{i1})) = \mathbf{A}_2 \ln(\mathbf{I} \cdot \boldsymbol{\pi}) = \mathbf{A}_2 \ln(\boldsymbol{\pi})$$

where the elements of $\mathbf{A}_2$ are simply the coefficients of the natural logarithms of the probabilities. Chapter 7 discusses an application of the logit function.

The researcher may also form more complex functions as shown by Forthofer and Koch (1973). Complex functions, which we analyze in Chapters 8 to 11, permit the researcher great flexibility in the types of analyses conducted. An example of the matrix expression used to produce a complex function, the rank correlation coefficient in Chapter 9, is

$$\mathbf{F} = \exp\{\mathbf{A}_4 \ln[\mathbf{A}_3 \exp\{\mathbf{A}_2 \ln(\mathbf{A}_1 \boldsymbol{\pi})\}]\}$$

This function may look complicated, but once the investigator determines what function of the probabilities to use, it is a matter of separating the function into its individual components, determining how each component is formed, and then linking the components together.

Once $\mathbf{F}$ is selected and created, the next step is to use the linear model approach to analyze the relationship between $\mathbf{F}$ and the factor variables used in the creation of the contingency table. To do this, however, we require knowledge of the variance-covariance matrix of $\mathbf{F}$—the topic of the next section.

THE VARIANCE OF A FUNCTION

The calculation of the variance-covariance matrix of $\mathbf{F}$ requires knowledge of the variance-covariance matrix of $\mathbf{p}$. Note that $\mathbf{p}$, the vector of the observed probabilities (relative frequencies), is arranged in the same order as $\boldsymbol{\pi}$. The vector $\mathbf{p}$ is an estimate of $\boldsymbol{\pi}$, and an element of $\mathbf{p}$, p_{ij}, is defined as

$$p_{ij} = \frac{n_{ij}}{n_{i\cdot}}$$

Since the data follow the multinomial distribution, we know that the variance of p_{ij} is

$$\text{var}(p_{ij}) = \frac{\pi_{ij}(1 - \pi_{ij})}{n_{i\cdot}}$$

Likewise from the multinomial distribution we know that the covariance of p_{ij} and p_{ik} is defined by

$$\text{cov}(p_{ij}, p_{ik}) = \frac{-\pi_{ij}\pi_{ik}}{n_{i\cdot}} \qquad j \neq k$$

The estimates of the variances and covariance terms are found by substituting p's for π's in the expressions above. We now use matrix notation to summarize these statements:

$$\underset{(r \times r)}{\mathbf{V}_i} = \frac{1}{n_{i\cdot}} \begin{bmatrix} p_{i1}(1 - p_{i1}) & -p_{i1}p_{i2} & \cdots & -p_{i1}p_{ir} \\ -p_{i1}p_{i2} & p_{i2}(1 - p_{i2}) & \cdots & -p_{i2}p_{ir} \\ \vdots & \vdots & & \vdots \\ -p_{i1}p_{ir} & -p_{i2}p_{ir} & \cdots & p_{ir}(1 - p_{ir}) \end{bmatrix}$$

The estimated variance-covariance matrix for all s subpopulations is

$$\underset{(rs \times rs)}{\mathbf{V}_p} = \begin{bmatrix} \mathbf{V}_1 & 0 & \cdots & 0 \\ 0 & \mathbf{V}_2 & \cdots & 0 \\ \vdots & \vdots & & \vdots \\ 0 & 0 & \cdots & \mathbf{V}_s \end{bmatrix}$$

The variance-covariance matrix $\mathbf{V}_p$ is a block-diagonal matrix with $r \times r$ matrices of zeros in the off-diagonal positions. This block-diagonal structure results from considering the subpopulations as being uncorrelated with one another; hence there are zero covariances between the subpopulations.

For a linear function, $\mathbf{F} = \mathbf{A}\pi$, and a linear combination of the logarithm of the probabilities, $\mathbf{F} = \mathbf{A} \ln(\pi)$, the $\mathbf{A}$ matrix is a matrix of constants. These constants are the coefficients of π or $\ln(\pi)$. The estimated variance-covariance matrix of the estimated function $\mathbf{F} = \mathbf{A}p$ is

$$\text{var}(\mathbf{F}) = \mathbf{V}_F = \mathbf{A}\mathbf{V}_p\mathbf{A}' \tag{3.1}$$

Note the similarity of this formula to the variance of a constant times a variable—that is, variance of cy, where c is a constant and y is a variable. In symbols this is

$$\text{var}(cy) = c^2 \text{ var}(y)$$
$$= c \text{ var}(y)c \qquad (3.2)$$

The formula for the estimated variance of $\mathbf{F} = \mathbf{Ap}$ shown in Equation (3.1) is a straightforward extension of this equation for matrices.

The estimated variance-covariance matrix for $\mathbf{F} = \mathbf{A} \ln(\mathbf{p})$ is

$$\text{var}(\mathbf{F}) = \mathbf{AD}^{-1}\mathbf{V}_p\mathbf{D}^{-1}\mathbf{A}' \qquad (3.3)$$

where $\mathbf{D}$ is an $rs \times rs$ diagonal matrix with the elements of $\mathbf{p}$ on the main diagonal. This formula again is an extension of the variance of a constant times a variable as shown in Equation (3.1). Forthofer and Koch (1973) and Landis and others (1976) provide formulas for the variance-covariance matrix of more complicated forms for $\mathbf{F}$.

The calculation of the variance-covariance matrix of $\mathbf{F}$ allows us now to use the linear model theory in the analysis of contingency tables, the subject of the next section.

THE LINEAR MODEL AND CATEGORICAL DATA

We will use the function of the probabilities as a dependent variable and attempt to determine if and how it is related to the independent or factor variables. We use the linear model approach to study the possible relationships by treating $\mathbf{F}$ as a linear function of the factors in the model. This relationship may be expressed as $\mathbf{F} = \mathbf{X}\boldsymbol{\beta}$, which is analogous to the analysis of variance. There is one important difference, however, between the analysis of variance and WLS categorical data analysis. In the analysis of variance we assume that the dependent variables have the same variance, but that assumption, also called the *assumption of homoscedasticity*, is unreasonable for the elements in $\mathbf{F}$. Instead of using least squares analysis to estimate $\boldsymbol{\beta}$ as in analysis of variance— that is, $\mathbf{b} = (\mathbf{X}'\mathbf{X})^{-1}\mathbf{X}'\mathbf{F}$—we use *weighted* least squares to find $\mathbf{b}$. The WLS estimator is

$$\mathbf{b} = (\mathbf{X}'\mathbf{V}_F^{-1}\mathbf{X})^{-1}\mathbf{X}'\mathbf{V}_F^{-1}\mathbf{F}$$

The weights are the elements of the inverse of the estimated variance-covariance matrix of $\mathbf{F}$. This procedure gives more weight to the elements in $\mathbf{F}$ that have smaller variances—that is, to the elements in which we have more confidence.

The variance-covariance matrix of $\mathbf{b}$ is

$$\text{var}(\mathbf{b}) = \mathbf{V}_b = (\mathbf{X}'\mathbf{V}_F^{-1}\mathbf{X})^{-1}$$

In the following material we use $\mathbf{X}$ to represent the matrix that connects $\mathbf{F}$ to the terms (the $\boldsymbol{\beta}$) in the model. We use X^2 to represent a statistic that asymptotically follows a central χ^2 distribution if the hypothesis being tested is true. This usage is consistent with the statistical literature. Note that the X^2 notation for the test statistic does not equal the square of the $\mathbf{X}$ matrix.

After obtaining estimates of the parameters in the model, we must assess whether the model agrees with the observed data. The test statistic for this determination is the goodness-of-fit X^2 written as

$$X^2_{\text{GOF}} = (\mathbf{F} - \mathbf{Xb})'\mathbf{V}_F^{-1}(\mathbf{F} - \mathbf{Xb})$$

and this has approximately a central chi-square distribution for large samples if the model fits. The degrees of freedom associated with X^2_{GOF} are equal to the number of rows in $\mathbf{F}$ minus the number of rows in $\boldsymbol{\beta}$. If the model does not fit, we may examine additional models suggested by the substantive theory.

If the model fits, we may proceed to determine which factors are related to the response variable by testing linear hypotheses about the $\boldsymbol{\beta}$. We express these linear hypotheses by H_0: $\mathbf{C}\boldsymbol{\beta} = \mathbf{0}$. Here the $\mathbf{C}$ matrix is used to form the desired linear combinations of the elements in $\boldsymbol{\beta}$ in exactly the same fashion that the $\mathbf{A}$ matrix is used to create the desired linear combinations of the elements of $\boldsymbol{\pi}$ when we create linear functions about $\boldsymbol{\pi}$—that is, $\mathbf{F} = \mathbf{A}\boldsymbol{\pi}$. The elements of $\mathbf{C}$ are the coefficients of the elements of $\boldsymbol{\beta}$ in the hypothesis we are testing.

The test statistic of the hypothesis H_0: $\mathbf{C}\boldsymbol{\beta} = \mathbf{0}$ is given by

$$X^2 = (\mathbf{Cb})'[\mathbf{C}(\mathbf{X}'\mathbf{V}_F^{-1}\mathbf{X})^{-1}\mathbf{C}']^{-1}\mathbf{Cb}$$

which has approximately a chi-square distribution in large samples if the hypothesis is true. The degrees of freedom associated with this test are equal to the number of rows in $\mathbf{C}$.

Once we are satisfied with the estimated model, we can also examine the predicted values of the function $\hat{\mathbf{F}} = \mathbf{XB}$ and its variance-covariance matrix

$$\text{var}(\hat{\mathbf{F}}) = \mathbf{V}_{\hat{F}} = \mathbf{X}[\mathbf{X}'\mathbf{V}_F^{-1}\mathbf{X}]^{-1}\mathbf{X}'$$

The $\hat{\mathbf{F}}$ represents an improved estimate of $\mathbf{F}$ because we used the entire data set in estimating $\hat{\mathbf{F}}$ whereas in calculating the ith element of $\mathbf{F}$ we used only the data from the ith subpopulation.

AN EXAMPLE OF WLS ANALYSIS

In the following example we demonstrate the application of these results to data from the 1964 Presidential Election Survey conducted by the University of Michigan Survey Research Center, as reported by Lehnen and Koch (1974c,

**Table 3.5   Attitudes Toward School Desegregation by Race and
Political Party Identification**

FACTOR VARIABLE		ATTITUDE TOWARD SCHOOL DESEGREGATION		Probability of Agreeing
Education	*Party Identification*	*Agree*	*Other*	
Less than high school	Democrat	304	225	0.575
	Independent	23	27	0.460
	Republican	49	79	0.383
High school or more	Democrat	287	197	0.593
	Independent	31	34	0.477
	Republican	134	158	0.459

Table 4). We analyze the data shown in Table 3.5 by treating the education by party identification row totals as fixed in advance. This analysis is called a *one-response and two-factor* situation.

We wish to determine whether the probability of agreeing is related to education and political party identification. The **A** matrix that generates the probability of agreeing is

$$\begin{bmatrix} 1 & 0 & 0 & 0 & 0 & 0 & 0 & 0 & 0 & 0 & 0 & 0 \\ 0 & 0 & 1 & 0 & 0 & 0 & 0 & 0 & 0 & 0 & 0 & 0 \\ 0 & 0 & 0 & 0 & 1 & 0 & 0 & 0 & 0 & 0 & 0 & 0 \\ 0 & 0 & 0 & 0 & 0 & 0 & 1 & 0 & 0 & 0 & 0 & 0 \\ 0 & 0 & 0 & 0 & 0 & 0 & 0 & 0 & 1 & 0 & 0 & 0 \\ 0 & 0 & 0 & 0 & 0 & 0 & 0 & 0 & 0 & 0 & 1 & 0 \end{bmatrix}$$

Now that **F** has been formed, we wish to relate it to education and political party. To represent the relationship we use the linear model $\mathbf{F} = \mathbf{X}\boldsymbol{\beta}$, where the elements of **X** indicate the factor level combination being considered. We can examine the education factor, the party identification factor, and an education by party interaction. In this example we directly examine the main effects of education and party identification. The $\boldsymbol{\beta}$ vector is

$$\begin{bmatrix} \text{mean term} \\ \text{differential} < \text{high school effect} \\ \text{differential Democrat effect} \\ \text{differential Independent effect} \end{bmatrix} = \begin{bmatrix} \beta_0 \\ \beta_1 \\ \beta_2 \\ \beta_3 \end{bmatrix}$$

This model has already been coded so that $\mathbf{X'X}$ will be nonsingular. The $\mathbf{X}$ matrix is

$$\mathbf{X} = \begin{bmatrix} 1 & 1 & 1 & 0 \\ 1 & 1 & 0 & 1 \\ 1 & 1 & -1 & -1 \\ 1 & -1 & 1 & 0 \\ 1 & -1 & 0 & 1 \\ 1 & -1 & -1 & -1 \end{bmatrix}$$

Any lack of fit in this model is due to the interaction of the education and political party variables. The goodness-of-fit test statistic is $X^2_{\text{GOF}} = 0.94$ with 2 degrees of freedom (six rows in $\mathbf{F}$ minus four rows in $\boldsymbol{\beta}$). This value is not significant for $\alpha = 0.25$. The parameter estimates and their estimated standard errors are

$$\mathbf{b} = \begin{bmatrix} 0.494 \\ -0.016 \\ 0.091 \\ -0.026 \end{bmatrix} \qquad SE(\mathbf{b}) = \begin{bmatrix} 0.018 \\ 0.013 \\ 0.020 \\ 0.032 \end{bmatrix}$$

Since the model fits, hypotheses about $\boldsymbol{\beta}$ can be tested; we will test them at $\alpha = 0.01$. The hypotheses of no significant effects due to education and political party membership with regard to the dependent variable will be tested:

$$H_0: \ \beta_1 = 0 \quad \text{and} \quad H_0: \begin{cases} \beta_2 = 0 \\ \beta_3 = 0 \end{cases}$$

The form of the test statistic is

$$X^2 = (\mathbf{Cb})'[\mathbf{C}(\mathbf{X'V}_F^{-1}\mathbf{X})^{-1}\mathbf{C'}]^{-1}\mathbf{Cb}$$

The $\mathbf{C}$ matrix for the education hypothesis is $\mathbf{C} = [0 \quad 1 \quad 0 \quad 0]$ and the value of X^2 is 1.587, which is less than $\chi^2_{1,0.99} = 6.63$. Therefore we fail to reject the hypothesis of no education effect.

The $\mathbf{C}$ matrix for the test of no significant party membership effects is

$$\mathbf{C} = \begin{bmatrix} 0 & 0 & 1 & 0 \\ 0 & 0 & 0 & 1 \end{bmatrix}$$

and $X^2 = 30.486$. Since this value is greater than 9.210 ($\chi^2_{2,0.99}$), we reject the hypothesis of no political party effects.

The results of this analysis suggest that the education level (less than high school versus at least a high school graduate) has no effect on whether the person is in favor of desegregation or not. Political party identification is highly related to attitude, however, with Democrats having a higher proportion of respondents strongly in favor of school desegregation (0.091 more than the average) while Independents are close to the overall average (-0.026) and Republicans have the lowest proportion (-0.065 below the average).

This example has demonstrated the use of functions—in this case a simple linear function of the probabilities—in model fitting and in testing hypotheses about parameters in the model. It is representative of the analyses in the chapters to follow regarding the creation of a function and its analysis by means of the linear model.

Now let us see how functions can lead to the *direct testing of hypotheses*. The two hypotheses we examine are two important hypotheses in the literature: the tests of *no additive association* and *no multiplicative association*.

ADDITIVE ASSOCIATION

In this section we demonstrate how a linear function can lead directly to an interesting hypothesis. We use the data presented in Table 3.3 and wish to determine whether there is a relationship between the passage of the HMO Act and the proportions of terminated and active. One way of examining this question is to test the hypothesis that the difference in the probabilities of terminated and active before the HMO Act is the same as the difference after the HMO Act.

The function of the probabilities that we use in the hypothesis is

$$(\pi_{11} - \pi_{12}) - (\pi_{21} - \pi_{22})$$

and the hypothesis itself is

$$H_0: \quad \pi_{11} - \pi_{12} - \pi_{21} + \pi_{22} = 0$$

This hypothesis is called the hypothesis of *no additive association*. If it is true, the distribution of the probabilities in each row is the same on a linear scale.

In this case we can simplify the form of the hypothesis because we know that $\pi_{i1} + \pi_{i2} = 1$ and this means that $\pi_{i2} = 1 - \pi_{i1}$. Therefore we have

$$H_0: \quad \pi_{11} - (1 - \pi_{11}) - \pi_{21} - (1 - \pi_{21}) = 0$$

or

$$H_0: \quad 2\pi_{11} - 1 - 2\pi_{21} + 1 = 0$$

which simplifies to

$$H_0: \quad \pi_{11} - \pi_{21} = 0$$

This is the hypothesis of *no additive association* based on the differences in the probabilities across the rows of a 2×2 contingency table.

MULTIPLICATIVE ASSOCIATION

In this section we examine a more complex function that also leads to an interesting hypothesis. The data, an aggregation of the data in Table 3.5, are shown in Table 3.6. One question in the survey dealt with attitudes toward school integration. We wish to determine whether attitude about this issue is independent of political party identification. In this example we consider both these variables as response variables because neither the attitude nor political party subtotals were fixed in advance. However, these data could also be viewed as a one-response and one-factor table if one wished.

Table 3.6   Attitudes Toward School Desegregation by Political Party Identification

Attitude:	AGREE			OTHER			
Party Identification:	Dem.	Ind.	Rep.	Dem.	Ind.	Rep.	Total
Frequencies	591	54	183	422	61	237	1548
Probabilities	π_{11}	π_{12}	π_{13}	π_{14}	π_{15}	π_{16}	1

Table 3.7 shows the same data as Table 3.6, but the information is now rearranged as a 3×2 table with a corresponding relabeling of the probabilities. Note that the probabilities in Table 3.7 are still treated as coming from a two-response, no-factor situation.

The question that is usually tested in two-way tables in introductory statistics courses is whether the row and column variables are independent of one another. This is expressed mathematically as

$$H_0: \quad \pi_{ij} = \pi_{i.}\pi_{.j} \quad \text{for } i = 1, 2, 3; j = 1, 2 \qquad (3.4)$$

Table 3.7   Rearrangement of Table 3.6

	FREQUENCIES		PROBABILITIES		
Party Identification	Agree	Other	Agree	Other	Total
Democrat	591	422	π_{11}	π_{12}	$\pi_{1\cdot}$
Independent	54	61	π_{21}	π_{22}	$\pi_{2\cdot}$
Republican	183	237	π_{31}	π_{32}	$\pi_{3\cdot}$
Total	—	—	$\pi_{\cdot 1}$	$\pi_{\cdot 2}$	1.00

There is an equivalent way of expressing this hypothesis, however. The first step involves writing out the hypothesis for the six combinations of i and j:

$$\pi_{11} = \pi_{1\cdot}\pi_{\cdot 1} \qquad \pi_{21} = \pi_{2\cdot}\pi_{\cdot 1} \qquad \pi_{31} = \pi_{3\cdot}\pi_{\cdot 1}$$

$$\pi_{12} = \pi_{1\cdot}\pi_{\cdot 2} \qquad \pi_{22} = \pi_{2\cdot}\pi_{\cdot 2} \qquad \pi_{32} = \pi_{3\cdot}\pi_{\cdot 2}$$

If we examine the ratio of π_{i1}/π_{i2}, we see that all three are equal to $\pi_{\cdot 1}/\pi_{\cdot 2}$ and hence the three are equal to one another. Therefore if the hypothesis H_0: $\pi_{ij} = \pi_{i\cdot}\pi_{\cdot j}$ is true, this implies that

$$H_0: \quad \frac{\pi_{i1}}{\pi_{i2}} = \frac{\pi_{j1}}{\pi_{j2}} \tag{3.5}$$

is also true.

If we can show that the hypothesis (3.5) implies hypothesis (3.4), we will have shown their equivalence. Assume that hypothesis 3.5 is true; that is,

$$\frac{\pi_{11}}{\pi_{12}} = \frac{\pi_{21}}{\pi_{22}} \quad \text{and} \quad \frac{\pi_{11}}{\pi_{12}} = \frac{\pi_{31}}{\pi_{32}}$$

Since we also know that the sum of these six probabilities is 1, we can express π_{22} as

$$\pi_{22} = 1 - \pi_{11} - \pi_{12} - \pi_{21} - \pi_{31} - \pi_{32}$$

Since $\pi_{11}/\pi_{12} = \pi_{21}/\pi_{22}$, this means that

$$\pi_{11} = \frac{\pi_{12}\pi_{21}}{1 - \pi_{11} - \pi_{12} - \pi_{21} - \pi_{31} - \pi_{32}}$$

or

$$\pi_{11}(1 - \pi_{11} - \pi_{12} - \pi_{21} - \pi_{31} - \pi_{32}) = \pi_{12}\pi_{21}$$

which gives

$$\pi_{11} = \pi_{11}(\pi_{11} + \pi_{12} + \pi_{21} + \pi_{31} + \pi_{32}) + \pi_{12}\pi_{21}$$

Now rearrange the terms:

$$\pi_{11} = \pi_{11}(\pi_{11} + \pi_{21} + \pi_{31})$$
$$+ \pi_{11}\pi_{12} + \pi_{11}\pi_{32} + \pi_{12}\pi_{21}$$

or

$$\pi_{11} = \pi_{11}\pi_{.1} + \pi_{12}(\pi_{11} + \pi_{21}) + \pi_{11}\pi_{32}$$

Since the second hypothesis is assumed to be true, we have

$$\frac{\pi_{11}}{\pi_{12}} = \frac{\pi_{31}}{\pi_{32}}$$

or

$$\pi_{11}\pi_{32} = \pi_{12}\pi_{31}$$

Therefore, if we replace $\pi_{11}\pi_{32}$ by $\pi_{12}\pi_{31}$ this yields

$$\pi_{11} = \pi_{11}\pi_{.1} + \pi_{12}(\pi_{11} + \pi_{21} + \pi_{31})$$

Since $\pi_{.1} = (\pi_{11} + \pi_{21} + \pi_{31})$, we simplify this expression to

$$\pi_{11} = \pi_{11}\pi_{.1} + \pi_{12}\pi_{.1} = \pi_{.1}(\pi_{11} + \pi_{12}) = \pi_{.1}\pi_{1.}.$$

Thus we have shown that

$$H_0: \quad \pi_{ij} = \pi_{i.}\pi_{.j}$$

is implied by

$$H_0: \quad \frac{\pi_{i1}}{\pi_{i2}} = \frac{\pi_{j1}}{\pi_{j2}}$$

The same steps can be used for the other π_{ij}. *Therefore the two forms of the hypotheses are equivalent.*

Examining the expression (3.5), we see that it says the ratio of the probabilities in one row is the same as the ratio in other rows. This hypothesis has been called the hypothesis of *no multiplicative association*. Note the similarity to the hypothesis of no additive association in the previous example. In additive association we consider *differences* whereas in multiplicative association we compare ratios. Note also that the hypothesis of no multiplicative association is the familiar hypothesis of independence of rows and columns in two-way tables.

The hypothesis of no multiplicative association

$$H_0: \quad \frac{\pi_{11}}{\pi_{12}} = \frac{\pi_{21}}{\pi_{22}} \quad \text{and} \quad \frac{\pi_{11}}{\pi_{12}} = \frac{\pi_{31}}{\pi_{32}}$$

or

$$H_0: \quad \frac{\pi_{11}\pi_{22}}{\pi_{12}\pi_{21}} = 1 \quad \text{and} \quad \frac{\pi_{11}\pi_{32}}{\pi_{12}\pi_{31}} = 1$$

involves a product of the probabilities, which is complicated. Additive functions are more tractable mathematically, and for that reason we are going to transform the hypothesis to an equivalent form by taking the logarithm of the equations for the hypothesis of no multiplicative association. The logarithm transformation converts multiplicative functions to additive functions in terms of logarithms. The hypothesis becomes

$$H_0: \quad \ln\left(\frac{\pi_{11}\pi_{22}}{\pi_{12}\pi_{21}}\right) = \ln(1)$$

$$\ln\left(\frac{\pi_{11}\pi_{32}}{\pi_{12}\pi_{31}}\right) = \ln(1)$$

or

$$H_0: \quad \ln(\pi_{11}) - \ln(\pi_{12}) - \ln(\pi_{21}) + \ln(\pi_{22}) = 0$$

$$\ln(\pi_{11}) - \ln(\pi_{12}) - \ln(\pi_{31}) + \ln(\pi_{32}) = 0$$

The left-hand side of these equations represents the functions of the probabilities. It allows us to determine whether the party identification and attitudes toward desegregation are independent—that is, are not associated in a multiplicative sense.

It is also possible to test the hypothesis of no additive association for these data. The functions that we create here are

$$\pi_{11} - \pi_{12} = \pi_{21} - \pi_{22}$$

and

$$\pi_{11} - \pi_{12} = \pi_{31} - \pi_{32}$$

The corresponding hypothesis is

$$H_0: \quad \pi_{11} - \pi_{12} - \pi_{21} + \pi_{22} = 0$$
$$\pi_{11} - \pi_{12} - \pi_{31} + \pi_{32} = 0$$

This looks similar to the hypothesis of no multiplicative association except that it is not expressed in terms of logarithms of the probabilities.

The choice of whether to test the additive or multiplicative form of the hypothesis of no association is discussed in Chapter 2. For commonly occurring events ($0.2 < \pi_{ij} < 0.8$), there should not be much difference observed in the substantive conclusions reached under either form of the hypothesis. For extreme events ($\pi_{ij} < 0.2$ or $\pi_{ij} > 0.8$), the multiplicative form of the hypothesis may be preferred as discussed in Chapter 2.

SUMMARY

In this chapter we introduced the contingency table and the WLS procedure for analyzing it. We also presented two types of association, additive and multiplicative, and demonstrated that the hypothesis of no multiplicative association is equivalent to the hypothesis of independence in two-way contingency tables.

These are the key topics covered in the chapter:

- *Contingency table:* Results from the cross-classification of a number of categorical variables.
- *Formation of functions:* The choice of function depends on the goals of the investigator; it may be simple or complex.
- V_F: Extension of $\text{var}(cy) = c \cdot \text{var}(y) \cdot c$.
- *Linear model approach:* $\mathbf{F} = \mathbf{X}\boldsymbol{\beta}$; the function formed is related in a linear manner to the factor variables; weighted by $\mathbf{V}_F^{-1}$.
- *Additive association:* Do the *differences* of the probabilities differ across the rows?
- *Multiplicative association:* Do the *ratios* of probabilities differ across the rows? Equivalent to independence in two-way tables.

The chapters in the remainder of the book consider problems in depth using the guidelines from Chapter 2 and the WLS procedure presented in this chapter.

EXERCISES

3.1.

The following data, summarized from a report by Lombard and Doering (1947) and subsequently analyzed by Dyke and Patterson (1952) and Cox and Snell (1968), relate knowledge of cancer to media exposure:

MEDIA EXPOSURE		KNOWLEDGE OF CANCER	
Newspapers	*Radio*	*Good*	*Poor*
Read	Listen	168	138
	Do not listen	310	357
Do not read	Listen	34	72
	Do not listen	156	494

Examine the relationship between knowledge of cancer and whether people read the newspaper, listen to the radio, or both, by considering the following questions:

    a. What is the appropriate **A** matrix that selects the probabilities of good knowledge?

    b. What is the appropriate design (**X**) matrix to represent the model that has a general mean plus the main effect of radio and newspaper? Assume that the effects are measured from their means.

    c. How many degrees of freedom are left over—that is, are associated with the error term (lack of fit)?

    d. Is it possible to include an interaction term in the model? How many degrees of freedom are associated with the radio by newspaper interaction?

3.2.

The accompanying table reports data for a hypothetical experiment involving the responses of 46 subjects to three different drugs. Alternative analyses of these data have been done by Grizzle, Starmer, and Koch (1969), Koch and Reinfurt (1971), and Bishop, Fienberg, and Holland (1975). The drugs are assumed to have no carryover effects, and the nature and severity of the condition being treated do not change over the course of drug testing.

DRUG			
A	*B*	*C*	*Number of Subjects*
F	F	F	6
F	F	U	16
F	U	F	2
F	U	U	4
U	F	F	2
U	F	U	4
U	U	F	6
U	U	U	6

F means favorable response; U means unfavorable response.

   a. What is the appropriate **A** matrix that will generate the vector of favorable responses for each of the three drugs?
   b. In the linear model $\mathbf{F} = \mathbf{X}\boldsymbol{\beta}$, where $\mathbf{X} = \mathbf{I}_3$, what is an appropriate **C** matrix in $H_0 : \mathbf{C}\boldsymbol{\beta} = \mathbf{0}$ to test the hypothesis that the favorable responses to the three drugs are equal?

3.3.

It is also possible to test the hypothesis of no three-way multiplicative association among the three drugs for the data from Exercise 3.2. The data may be reformulated as:

		Drug A	
Drug B	*Drug C*	*F*	*U*
F	F	π_1	π_2
F	U	π_3	π_4
U	F	π_5	π_6
U	U	π_7	π_8

The association of drugs A and C within the F level of B can be represented by

$$\frac{\pi_1 \pi_4}{\pi_2 \pi_3}$$

The association of drugs A and C within the U level of B is

$$\frac{\pi_5 \pi_8}{\pi_6 \pi_7}$$

Hence if drug B does not affect the relationship between drugs A and C, these two measures of association between A and C will be equal. The hypothesis of no multiplicative association among the three drugs is

$$H_0: \quad \frac{\pi_1 \pi_4}{\pi_2 \pi_3} = \frac{\pi_5 \pi_8}{\pi_6 \pi_7}$$

and this can be expressed as

$$H_0: \quad \frac{\pi_1 \pi_4}{\pi_2 \pi_3} \frac{\pi_6 \pi_7}{\pi_5 \pi_8} = 1$$

The hypothesis of no three-way multiplicative association can be expressed in terms of logarithms of π—that is,

$$H_0: \quad \mathbf{A} \ln(\pi) = \mathbf{0}$$

What is the proper $\mathbf{A}$ matrix that generates this hypothesis?

PART II

Simple Applications of the WLS Approach

4
One Response and
Two Factor Variables

This chapter details a step-by-step application of the WLS procedure on a small problem posing few complexities. The steps presented here will serve as a guide for the more complicated analyses presented in later chapters and thus should be mastered before proceeding further.

The problem is to determine what conditions affect sentencing of persons charged with the commission of a property crime. The data come from a study conducted in Charlotte-Mecklenburg County, North Carolina, to learn about the operations of the criminal court there and to examine the impact of reforms proposed by the American Bar Association (see Clarke and Koch, 1976; Clarke, Freeman, and Koch, 1976.) The example used here is taken from an analysis presented by Clarke and Koch (1976) that examined how certain factors—the offense charged, the defendant's prior record, race, sex, and income—affected the chance of receiving a prison sentence.

ANALYZING SENTENCES IN CRIMINAL COURT

To simplify this example, we have selected two important variables related to a defendant's receiving a prison sentence. In the first part of this chapter we present an analysis of three variables—the relationship among the offense charged, the defendant's prior record, and the court's disposition of the case. The next chapter, which extends the analysis, includes other relevant variables identified by Clarke and Koch (1976).

The data from the Charlotte-Mecklenburg study can be viewed from several perspectives. The 798 cases are not a representative sample of data from all criminal courts across the United States. Rather, they reflect a legal process that is limited by time and the place of selection. As we discussed in Chapter 2, tests of significance in this context can best be treated as guides pointing to important sources of variation. Thus we may use these tests as general indicators—first, to select the variables most likely to influence court outcomes, and second, to eliminate unimportant sources of variation. One may expect variation in the data because a sample drawn at different times and places will no doubt differ from the one analyzed here. The recording and classification of criminal offenses is not error free and thus introduces additional variation; therefore the model must be treated as tentative. It is impossible to know which sources of variation are real—in the sense that they would be present in all samples at all times and in all places—and which ones are unique to this study. Only replication of these results will help answer the questions about the true model. In sum, the model-fitting exercise in this chapter seeks to find an economical means of explaining why the likelihood of prison convictions varies. The final model is limited in generality by time and place and by the subset of predictors selected from a theoretically infinite set of possible predictor variables.

DEFINING THE VARIABLES

The question to be answered is whether the existence of a prior criminal record and the nature of the offense charged affect the chance of receiving a prison sentence. The data in Table 4.1 are for 798 criminal cases involving defendants charged with a property offense. Each case is classified according to the following

Table 4.1   Disposition of Criminal Cases by Prior Arrest Record and Type of Offense

		DISPOSITION	
Prior Arrest Record	*Type of Offense*	*No Prison*	*Prison*
None	Nonresidential burglary	38	17
None	Other property	244	21
Some	Nonresidential burglary	67	42
Some	Other property	302	67

Source: Adapted from Clarke and Koch (1976, table 6).

three dichotomous variables:

1. Disposition of case (response variable): defendant received prison sentence/defendant received other sentence
2. Defendant's prior arrest record (factor variable): some prior arrest record/no prior arrest record
3. Type of property offense charged (factor variable): defendant charged with nonresidential burglary/defendant charged with other property offense

These three variables produce a three-dimensional table of size $2 \times 2 \times 2$. This table can be viewed as having one response and two factors.

The analysis of categorical data begins with specifying the elements of the linear model:

$$\mathbf{F} = \mathbf{X}\boldsymbol{\beta}$$

where $\mathbf{F}$ is a vector of response functions, $\mathbf{X}$ is a design matrix specified by the analyst, and $\boldsymbol{\beta}$ is the vector of parameters for this problem.

THE RESPONSE FUNCTION

The problem is to analyze the probability of receiving a prison sentence under each of four conditions:

1. The defendant had no prior record and was charged with a nonresidential burglary.
2. The defendant had no prior record and was charged with some other property crime.
3. The defendant had some prior record and was charged with a nonresidential burglary.
4. The defendant had some prior record and was charged with some other property crime.

A *response function* is a mathematical expression based on the categories of the dependent variable. For this analysis, the response function is the proportion of defendants that received a prison sentence:

$$\text{Probability of prison} = \frac{\text{number given a prison sentence}}{\substack{\text{number given a prison sentence} + \\ \text{number who did not receive a} \\ \text{prison sentence}}}$$

This estimated probability is calculated for each of the four groups specified above. The observed proportions, which estimate this probability of prison,

**Table 4.2 Observed Cell Proportions Based on Frequency Data
from Table 4.1**

| Prior Arrest Record | Type of Offense | DISPOSITION | |
		No Prison	*Prison*
None	Nonresidential burglary	$\dfrac{38}{38 + 17} = 0.691$	$\dfrac{17}{38 + 17} = 0.309$
None	Other property	$\dfrac{244}{244 + 21} = 0.921$	$\dfrac{21}{244 + 21} = 0.079$
Some	Nonresidential burglary	$\dfrac{67}{67 + 42} = 0.615$	$\dfrac{42}{67 + 42} = 0.385$
Some	Other property	$\dfrac{302}{302 + 67} = 0.818$	$\dfrac{67}{302 + 67} = 0.182$

are defined in Table 4.2. The *response vector* for this problem—its elements are the estimated probability of prison for the four conditions—is

$$\mathop{\mathbf{F}}_{(4 \times 1)} = \begin{bmatrix} 0.309 \\ 0.079 \\ 0.385 \\ 0.182 \end{bmatrix} \tag{4.2}$$

In Table 4.2 the rows represent *subpopulations* and are defined by the combinations of two factor variables. Any response function selected will be based on the cell proportions, under the restriction that the cell proportions sum to 1.00 for each subpopulation defined in the frequency table.

SPECIFYING THE MODEL

We now have specified the following linear model:

$$\mathbf{F} = \mathbf{X}\boldsymbol{\beta}$$

or

$$\begin{bmatrix} 0.309 \\ 0.079 \\ 0.385 \\ 0.182 \end{bmatrix} = \mathop{\mathbf{X}}_{(4 \times t)} \mathop{\boldsymbol{\beta}}_{(t \times 1)} \tag{4.3}$$

To estimate β we must specify the design matrix X, where X defines the linear model that we are interested in fitting. The choice of X is almost limitless, but some design matrices X are more suited to our purposes than others, in the sense that they produce a β that permits an interpretation of how prior arrest record and offense charged affect a defendant's chance of receiving a prison sentence.

Let us define X such that here are three *effects* to analyze:

$$\beta_0 = \text{average proportion of prison sentences}$$

$$\beta_1 = \text{differential "prior record" effect}$$

$$\beta_2 = \text{differential "type of offense" effect}$$

For the first two subpopulations, β_1 is added to β_0; for the last two, it is subtracted. These subpopulations correspond to the "none" and "some" prior arrest history subpopulations. The parameter β_2 acts similarly—it is added for the first and third subpopulations and subtracted for the second and fourth (the "nonresidential burglary" and "other property" subpopulations). These requirements produce the following linear model:

$$\begin{bmatrix} 0.309 \\ 0.079 \\ 0.385 \\ 0.182 \end{bmatrix} = \begin{bmatrix} 1 & 1 & 1 \\ 1 & 1 & -1 \\ 1 & -1 & 1 \\ 1 & -1 & -1 \end{bmatrix} \begin{bmatrix} \beta_0 \\ \beta_1 \\ \beta_2 \end{bmatrix} \tag{4.4}$$

The choice of this particular X permits a direct interpretation of the parameters. The estimated general mean, b_0, is an estimate of the probability of receiving a prison sentence, regardless of prior arrest history and offense charged; b_1 is an estimate of the differential effect of a prior arrest history; b_2 is an estimate of the differential effect of a nonresidential property offense.

ESTIMATING PARAMETERS

As shown in Chapter 3, the solution to the linear model may be found by

$$b = (X'V_F^{-1}X)^{-1}X'V_F^{-1}F \tag{4.5}$$

where F and X have been given above and V_F^{-1} is the inverse of the variance-covariance matrix of the response functions F. The estimated variance-covariance matrix V_F in this problem is

$$V_F = \begin{bmatrix} 0.00388 & 0.00000 & 0.00000 & 0.00000 \\ 0.00000 & 0.00028 & 0.00000 & 0.00000 \\ 0.00000 & 0.00000 & 0.00217 & 0.00000 \\ 0.00000 & 0.00000 & 0.00000 & 0.00040 \end{bmatrix} \tag{4.6}$$

and its inverse is

$$V_F^{-1} = \begin{bmatrix} 257.7 & 0.0 & 0.0 & 0.0 \\ 0.0 & 3571.4 & 0.0 & 0.0 \\ 0.0 & 0.0 & 460.8 & 0.0 \\ 0.0 & 0.0 & 0.0 & 2500.0 \end{bmatrix} \qquad (4.7)$$

Solving for **b** using expressions (4.2) and (4.5) to (4.7), we obtain

$$\mathbf{b} = \begin{bmatrix} b_0 \\ b_1 \\ b_2 \end{bmatrix} = \begin{bmatrix} 0.237 \\ -0.050 \\ 0.107 \end{bmatrix} \qquad (4.8)$$

EVALUATING THE MODEL'S FIT

Before we interpret **b** and draw some conclusions about the disposition of criminal cases in North Carolina, we need to evaluate critically whether $\mathbf{F} = \mathbf{X}\boldsymbol{\beta}$ is a suitable model for the data. One criterion for the evaluation is how well the model predicts the observed **F**—that is, the differences between the observed proportions and the proportions predicted from the fitted model. We examine this criterion below.

The predicted values for the fitted model are defined as

$$\hat{\mathbf{F}} = \mathbf{X}\mathbf{b}$$

In algebraic terms, this expression is

$$\begin{aligned} \hat{f}_1 &= b_0 + b_1 + b_2 \\ \hat{f}_2 &= b_0 + b_1 - b_2 \\ \hat{f}_3 &= b_0 - b_1 + b_2 \\ \hat{f}_4 &= b_0 - b_1 - b_2 \end{aligned} \qquad (4.9)$$

Substituting the values of **b** we have

$$\hat{\mathbf{F}} = \begin{bmatrix} 0.237 + (-0.050) + 0.107 \\ 0.237 + (-0.050) - 0.107 \\ 0.237 - (-0.050) + 0.107 \\ 0.237 - (-0.050) - 0.107 \end{bmatrix} = \begin{bmatrix} 0.294 \\ 0.080 \\ 0.394 \\ 0.180 \end{bmatrix} \qquad (4.10)$$

The errors in prediction based on $\mathbf{X}\boldsymbol{\beta}$ are defined as

$$e_i = f_i - \hat{f}_i \qquad (4.11)$$

and are shown in Table 4.3.

**Table 4.3 Observed and Predicted Proportions
and Residuals[a]**

Prior Arrest Record	Type of Offense	Observed f	Predicted $\hat{f}$	Residual e
None	Nonresidential burglary	0.309	0.294	0.015
None	Other property	0.079	0.080	−0.001
Some	Nonresidential burglary	0.385	0.394	−0.009
Some	Other property	0.182	0.180	0.002

Note: $X^2_{\text{GOF}} = 0.1011$; DF $= 1$.
[a] Based on model $\mathbf{F} = \mathbf{X}\boldsymbol{\beta}$ defined in Equation (4.4).

The errors or residuals based on this model are acceptable in the sense that they appear to be of the same magnitude—that is to say, the error in prediction is not unusually large for one subpopulation and small for the remainder. Rather, the errors appear to be distributed in an unpredictable manner. Table 4.3 summarizes the observed, predicted, and residual values for this analysis. For more on residual analysis the interested reader may see Draper and Smith (1966), Anscombe (1961), and Cox and Snell (1968).

We may gain an additional test of the model by hypothesizing that the errors are the result of chance variation in the data. This is a goodness-of-fit test of the model $\mathbf{F} = \mathbf{X}\boldsymbol{\beta}$. The goodness-of-fit test statistic X^2_{GOF} is defined in Chapter 3 and is distributed as a chi-square, when the sample size of the sub-population is sufficiently large, with DF $= u - t$. The term u is the number of functions analyzed; for this problem, $u = 4$. The term t is the number of parameters estimated and $t = 3$ in this example.

What is the goodness-of-fit test telling us about the model? Observe that we have exercised a type of statistical parsimony by predicting the four observed f_i based on a model having three parameters. The economy here is in obtaining four numbers for the price of three. In more complex problems where the number of functions fitted is large, the importance of this economy will be evident. In this example, let us concentrate on considering what the extra degree of freedom omitted from the model represents. In substantive terms it is the interaction between the differential prior record effect β_1 and the differential offense charged effect β_2. If it had been included, it could be defined as $\beta_3 = \beta_1\beta_2$ and our model would have been

$$\mathbf{F} = \begin{bmatrix} 1 & 1 & 1 & 1 \\ 1 & 1 & -1 & -1 \\ 1 & -1 & 1 & -1 \\ 1 & -1 & -1 & 1 \end{bmatrix} \begin{bmatrix} \beta_0 \\ \beta_1 \\ \beta_2 \\ \beta_3 \end{bmatrix}$$

Since there is only 1 degree of freedom omitted from $\mathbf{X}\boldsymbol{\beta}$, the goodness-of-fit test *in this application* is also a test for this interaction.

The chi-square goodness-of-fit test statistic $X^2_{\text{GOF}} = 0.1011$ is not significant at the 0.25 level since it is less than the value of $\chi^2_{1,0.75} = 1.32$. We do not reject the hypothesis that the model fits. Hence the predicted value of $\mathbf{F}$ associated with model $\mathbf{F} = \mathbf{X}\boldsymbol{\beta}$ is not statistically different from the observed value of $\mathbf{F}$. The model predicts the observed data and therefore it is appropriate to test additional hypotheses about the effects β_1 and β_2.

TESTING HYPOTHESES

Since the model has met our criteria for a satisfactory fit, the next step is to test hypotheses about the parameters under the assumption that the model is correct. The general procedure for testing hypotheses, $H_0: \mathbf{C}\boldsymbol{\beta} = \mathbf{0}$, is to define a matrix $\mathbf{C}$ and compute X^2 where

$$X^2 = (\mathbf{Cb})'[\mathbf{C}(\mathbf{X}'\mathbf{V}_F^{-1}\mathbf{X})^{-1}\mathbf{C}']^{-1}\mathbf{Cb}$$

The test statistic, X^2, follows approximately the chi-square distribution with c degrees of freedom in large samples if the hypothesis is true. Here c is the number of independent rows in $\mathbf{C}$. Two specific hypotheses interest us:

1. Prior record has no effect on the chance of receiving a prison sentence—that is, H_{01}: $\beta_1 = 0$.
2. The property offense charged has no effect on the chances of receiving a prison sentence—that is, H_{02}: $\beta_2 = 0$.

The tests of the specific hypotheses under $\mathbf{X}\boldsymbol{\beta}$ described above are generated by choosing

$$\mathbf{C}_1 = [0 \quad 1 \quad 0] \quad \text{and} \quad \mathbf{C}_2 = [0 \quad 0 \quad 1]$$

The test statistic for each hypothesis is

$$H_{01}: \quad \beta_1 = 0 \quad X^2_1 = 16.30 \quad (p < 0.05)$$
$$H_{02}: \quad \beta_2 = 0 \quad X^2_1 = 28.72 \quad (p < 0.05)$$

Based on these test statistics, we must reject the hypotheses that the parameters β_1 and β_2 are equal to zero ($\alpha = 0.05$).

INTERPRETING THE MODEL

Now that we have determined that the model fits the data and that parameters β_1 and β_2 are significantly different from zero, what can we conclude about court dispositions in North Carolina? We may say that this sample of cases shows that the average chance of receiving a prison sentence regardless of prior arrest record and type of offense charged is 0.237, or 23.7 percent (because $b_0 = 0.237$). If one has a prior arrest record, the chances of a prison sentence are increased by 0.05, or 5 percent ($b_1 = -0.05$). If one has no prior record, the chance is reduced by 5 percent. The property offense charged is also an important influence determining sentence. Defendants charged with non-residential burglaries have a 10.7 percent higher probability ($b_2 = 0.107$) of receiving a prison sentence, whereas those charged with other property crimes (mostly larceny) have their probability reduced by 10.7 percent. The two effects do not interact; hence defendants with prior arrest records and charged with nonresidential burglaries are most likely to receive prison sentences. Their chance of going to prison is 39.4 percent and, correspondingly, defendants without a prior record and charged with some other property offense are least likely to receive a prison sentence. The observed and predicted **F** and the residuals are summarized in Table 4.3.

SUMMARY

In this chapter we analyzed the effects of two conditions—prior record and offense charged—on the probability of receiving a prison sentence. To determine the magnitude and direction of these effects, we defined a response function—the proportion of defendants receiving a prison sentence—and fitted a linear model by means of weighted least squares procedures. After evaluating the model for lack of fit and finding it acceptable, we tested specific hypotheses about the effects of the factors on the response function. We found that both having a prior record and being tried for a nonresidential burglary increased the likelihood of receiving a prison sentence.

These are the key points covered in the chapter:

- *The problem:* Analysis of one response (probability of a prison sentence) and two factor variables (prior prison record and type of offense committed).
- *Variables:* Response variable (prison sentencing) and factor variables (prior prison record and type of crime).
- *Response function:* Probability of a prison sentence.
- *Model:* Definition of **X** and β.
- *Parameter estimation:* $\mathbf{b} = (\mathbf{X}'\mathbf{V}_F^{-1}\mathbf{X})^{-1}\mathbf{X}'\mathbf{V}_F^{-1}\mathbf{F}$.
- *Fit of the model:* $X_{\text{GOF}}^2 = 0.1011$; the model fits; error = observed − predicted.
- *Hypothesis:* Prior prison record is important, as is type of offense at $\alpha = 0.05$ level.

- *Interpretation:* Higher probability of a prison sentence is associated with prior arrest record and nonresidential burglary offenses.

The next chapter considers the same data set, but it includes an additional factor variable and expands the offense categories.

EXERCISES

4.1.

Duplicate by hand the analysis presented in this chapter by completing the following steps.

a. Compute the (8×1) vector $\mathbf{p}$ of observed proportions from the frequency data presented in Table 4.1. Given that $\mathbf{p}_1$ is the column vector of observed proportions for the first subpopulation $\begin{bmatrix} 0.691 \\ 0.309 \end{bmatrix}$, then

$$\mathbf{p}_{(8 \times 1)} = \begin{bmatrix} \mathbf{p}_1 \\ \mathbf{p}_2 \\ \mathbf{p}_3 \\ \mathbf{p}_4 \end{bmatrix}$$

b. Compute $\mathbf{V}_p$, the estimated variance-covariance matrix associated with $\mathbf{p}$, by using Equations (3.1) to (3.3) in Chapter 3.

c. Using the linear operator matrix

$$\mathbf{A} = \begin{bmatrix} 0 & 1 & 0 & 0 & 0 & 0 & 0 & 0 \\ 0 & 0 & 0 & 1 & 0 & 0 & 0 & 0 \\ 0 & 0 & 0 & 0 & 0 & 1 & 0 & 0 \\ 0 & 0 & 0 & 0 & 0 & 0 & 0 & 1 \end{bmatrix}$$

compute the (4×1) vector of observed proportions $\mathbf{F} = \mathbf{Ap}$. Compare your answer to Expression 4.2.

d. Compute the estimated variance-covariance matrix $\mathbf{V}_F$ for the vector $\mathbf{F}$ using the expression $\mathbf{V}_F = \mathbf{AV}_p\mathbf{A}'$. Compare your answer with Expression 4.6.

e. Find the inverse $\mathbf{V}_F^{-1}$. This matrix will be used in the following computations. (*Note:* Compute this matrix by hand or use a suitable computer program.)

f. At this point we have defined the vector $\mathbf{F}$—the observed proportions receiving a prison sentence for each subpopulation, based on the frequency data presented in Table 4.1—and computed $\mathbf{V}_F$, the observed variance-covariance matrix. Note the role that the linear operator matrix $\mathbf{A}$ has played in the computation of $\mathbf{F}$ and $\mathbf{V}_F$. The matrix $\mathbf{A}$ is defined according to the needs of the analyst. Had the analysis focused on the proportion of defendants receiving another sentence, the appropriate **A** matrix would have been

$$\mathbf{A} = \begin{bmatrix} 1 & 0 & 0 & 0 & 0 & 0 & 0 & 0 \\ 0 & 0 & 1 & 0 & 0 & 0 & 0 & 0 \\ 0 & 0 & 0 & 0 & 1 & 0 & 0 & 0 \\ 0 & 0 & 0 & 0 & 0 & 0 & 1 & 0 \end{bmatrix}$$

The next task is to define a suitable linear model $\mathbf{F} = \mathbf{X}\boldsymbol{\beta}$, compute the estimated parameter vector **b**, and evaluate its fit by using a goodness-of-fit statistic X_{GOF}^2 when using the design matrix

$$\mathbf{X} = \begin{bmatrix} 1 & 1 & 1 \\ 1 & 1 & -1 \\ 1 & -1 & 1 \\ 1 & -1 & -1 \end{bmatrix}$$

Interpret each b_i in the model.

g. Compute the (3×1) vector of estimated parameters **b** by using the expression

$$\mathbf{b} = (\mathbf{X}'\mathbf{V}_F^{-1}\mathbf{X})^{-1}\mathbf{X}'\mathbf{V}_F^{-1}\mathbf{F}$$

h. Find the estimated variance-covariance matrix $\mathbf{V}_b$ of **b**, where

$$\mathbf{V}_b = (\mathbf{X}'\mathbf{V}_F^{-1}\mathbf{X})^{-1}$$

i. Calculate the goodness-of-fit statistic

$$X_{\text{GOF}}^2 = (\mathbf{F} - \mathbf{Xb})'\mathbf{V}_F^{-1}(\mathbf{F} - \mathbf{Xb})$$

Is this statistic significant at $\alpha = 0.25$? Explain the meaning of this statistic.

j. Finally, test the hypothesis $H_0: \ \mathbf{C}\boldsymbol{\beta} = \mathbf{0}$, where $\mathbf{C} = \begin{bmatrix} 0 & 1 & 0 \end{bmatrix}$. The test statistic is

$$X^2 = (\mathbf{Cb})'[\mathbf{C}(\mathbf{X}'\mathbf{V}_F^{-1}\mathbf{X})^{-1}\mathbf{C}']^{-1}\mathbf{Cb}$$

Interpret this test statistic.

4.2.

Duplicate the analysis of this chapter by using the GENCAT program (Appendix D) and the computer input given at the end of the chapter.

a. What is the role of the **A** matrix?
b. What function does the **X** matrix play?

c. Why is more than one **C** matrix used?
d. Examine the computer output and find the following:

  1. The observed frequencies (Table 4.1)
  2. The observed proportion vector **p**
  3. The vector of functions **F**
  4. The estimated variance-covariance matrix V_F
  5. The vector of the estimated parameters **b**
  6. The estimated variance-covariance matrix V_b
  7. The predicted function vector $\hat{F}$
  8. The vector of residuals ($F - Xb$)
  9. The goodness-of-fit statistic X^2_{GOF}
  10. The test statistic for $C = \begin{bmatrix} 0 & 1 & 0 \end{bmatrix}$

4.3.

Table 4.4 is adapted from a study of attitudes toward conflict resolution among party activists in Berne Canton, Switzerland, conducted by Steiner and Lehnen (1974). The investigators wished to determine whether a respondent's prior leadership experience in canton or federal government, the Swiss Free Democratic Party (FDP), or an interest group (trade group, union, and the like) affected opinions regarding the criterion for making political decisions. The statement was: "It would be better if there were more clear decisions instead of compromises in Switzerland." The authors hypothesized that prior leadership

Table 4.4   Responses of 201 FDP Members from Berne Canton, Switzerland

Government	Party	Interest Group	Disagree	Agree	Total
Yes	Yes	Yes	14	19	33
Yes	Yes	No	18	9	27
Yes	No	Yes	4	6	10
Yes	No	No	23	13	36
No	Yes	Yes	3	7	10
No	Yes	No	14	9	23
No	No	Yes	5	6	11
No	No	No	31	20	51

PRIOR LEADERSHIP EXPERIENCE / RESPONSE

experience would be positively associated with attitudes supporting consensus (compromise) decision making.

a. How many independent variables (factors) and dependent variables (responses) are presented in this problem? How many subpopulations are defined by the factor combinations?

b. A "disagree" response to the statement "It would be better if there were more clear decisions instead of compromises in Switzerland" was interpreted as an indication of support for consensus (compromise) decision making. Compute by hand the observed proportions in each of the eight subpopulations favoring and opposing compromise decision making. Form a (16×1) column vector of observed proportions $\mathbf{p}$ and construct an (8×16) matrix $\mathbf{A}$ such that $\mathbf{F} = \mathbf{Ap}$ defines an (8×1) vector of functions $\mathbf{F}$, the proportions disagreeing for each of the eight subpopulations.

c. Write a system of linear equations $\mathbf{F} = \mathbf{X\beta}$ that expresses $\mathbf{F}$ as a function of the general mean, the main effects of government, party, and interest group leadership experience, and the interactions government $\times$ party, government $\times$ interest group, and party $\times$ interest group. This exercise is a prelude for Chapter 5. Write the equations using both differential effect and dummy coding (see Appendix B for a discussion of interaction).

d. Construct a suitable design matrix $\mathbf{X}$ for each system of equations.

e. Fit each model and interpret the goodness-of-fit tests under each. Examine the residuals for each model. Do the models adequately fit the data? Explain your conclusions.

f. Interpret the meaning of the general mean estimate for each model. Why do the models differ for the differential effect and dummy coding?

g. Test the statistical significance ($\alpha = 0.01$) of the three main effects and the three first-order interaction effects under the hypothesis H_0: $\mathbf{C\beta} = 0$. Write the $\mathbf{C}$ matrix for each hypothesis test. Interpret each test and write a sentence summarizing your substantive conclusion. Complete the following table, which summarizes the chi-square statistics for the six hypotheses under each model:

Effect	Differential Effect Coding	Dummy Coding
Government		
Party		
Interest group		
Government $\times$ party		
Government $\times$ interest group		
Party $\times$ interest group		

h. Using the information derived from the hypothesis tests conducted in (part g), fit a reduced model, evaluate the goodness of fit, and test the significance of the model parameters.

i. What substantive conclusion can you draw about the relationship between leadership experience and attitudes toward decision making? Write a paragraph discussing your conclusions. How far may one generalize these results—that is, to what population do your conclusions apply? Is the sample sufficient to support your conclusions?

APPENDIX: GENCAT INPUT

The following listing is for the GENCAT control cards that reproduce the analysis described in this chapter.

```
        5     1     1                    CHAPTER 4: ANALYSIS OF NC COURT DATA
        4     2
38.         17.
244.        21.
67.         42.
302.        67.
        1     2     4     1            (2F2.0)
0001
        7     1     3                  (4F2.0)           MAIN EFFECTS MODEL
01010101
0101-1-1
01-101-1
        8     1     1                  (3F2.0)           TEST: B1 = 0
0001
        8     1     1                  (3F2.0)           TEST: B2 = 0
000001
```

5

Interaction Among Factor Variables

In the last chapter we analyzed data on criminal court dispositions that could be explained by the main effects of the two factor variables. In this chapter we introduce more complexity into the analysis by adding another variable and another level to the type of offense variable. Because of these changes, a main-effects model will no longer fit the data and a model including *interactions* among the factors will be required. Besides introducing the modeling of interaction in this chapter, we also consider a new type of model—the *modular* model—in our attempt to characterize the variation in the data.

ADDING COMPLEXITY

In Chapter 4 we considered a simplified table derived from data on criminal court dispositions collected in Charlotte-Mecklenburg County, North Carolina. The larger study from which these data came examined the combined effects of sex, age, race, income, employment status, promptness of arrest, offense charged at time of arrest, and prior arrest record as factors determining the disposition of the case (Clarke and Koch, 1976). The authors of this study found that two variables, offense charged and prior arrest record, were the strongest determinants among the eight measured variables of a defendant receiving a prison sentence. Two other variables, the defendant's income and promptness of arrest (which the authors used as a surrogate for strength of evidence against the defendant), also appeared to play a lesser but statistically significant role in explaining the court's disposition. The remaining variables were not statistically related to the outcome of the case.

55

We will use three predictor variables—prior arrest record, offense charged, and defendant's income—to analyze variation in the likelihood of receiving a prison sentence. The variables are defined as:

1. Disposition of case (response variable): defendant received prison sentence/ defendant received other sentence
2. Defendant's prior arrest record (factor variable): some prior arrest record/no prior arrest record
3. Type of property offense charged (factor variable): defendant charged with nonresidential burglary/defendant charged with residential burglary/defendant charged with larceny-theft
4. Defendant's income (factor variable): high/low

The data for these variables are shown in Table 5.1 along with the observed proportion receiving a prison sentence for each of the 12 subpopulations. Table 5.1 is similar in structure to Table 4.1, except for the introduction of the income variable and the classification of offense into three levels instead of two. As a result of these changes, there are now 12 subpopulations.

Table 5.1   Disposition of Criminal Cases by Prior Arrest Record, Type of Offense, and Defendant's Income

| Prior Arrest Record | Type of Offense | Defendant's Income | DISPOSITION | | Observed Proportion Receiving Prison Sentence |
			No Prison	Prison	
None	Nonresidential burglary	Low	17	13	0.433
None	Nonresidential burglary	High	21	4	0.160
None	Residential burglary	Low	25	3	0.107
None	Residential burglary	High	23	2	0.080
None	Larceny	Low	77	7	0.083
None	Larceny	High	119	9	0.070
Some	Nonresidential burglary	Low	36	27	0.429
Some	Nonresidential burglary	High	31	15	0.326
Some	Residential burglary	Low	56	25	0.309
Some	Residential burglary	High	36	5	0.122
Some	Larceny	Low	119	29	0.196
Some	Larceny	High	91	8	0.081

Source: Adapted from Clarke and Koch (1976, table 6).

MODELING INTERACTION

Interaction means that there is variation in the dependent variable associated with two or more independent variables working together. In this chapter, for example, we will see that the combination of offense charged and defendant's income determines the degree to which another independent variable, prior record, affects the sentence received. Interaction is said to be present because the effect of prior record on sentence is not the same for all combinations of offense and income.

There are several means by which we can incorporate these substantive notions into the model. In the following pages we model interaction by using the product of effects. This approach is discussed in Appendix B in the context of analysis of variance. We also introduce a *modular* approach to modeling interaction—an approach that provides a readily interpretable mathematical model for characterizing many substantive formulations of interaction. Ideally, we desire to fit models that are simple and easy to interpret—such as *main effects models*, which contain no interaction terms. Unfortunately, main-effects models do not provide an adequate fit of most complicated problems. Certainly this is the case with the data presented in Table 5.1. The constant and main effects in this example may be defined as

$$
\begin{bmatrix}
\text{Constant} \\
\text{Differential effect of no prior arrest} \\
\text{Differential effect of nonresidential burglary} \\
\text{Differential effect of residential burglary} \\
\text{Differential effect of low income}
\end{bmatrix}
=
\begin{bmatrix}
\beta_0 \\
\beta_1 \\
\beta_2 \\
\beta_3 \\
\beta_4
\end{bmatrix}
$$

The main effects model is

$$
\begin{bmatrix}
0.433 \\
0.160 \\
0.107 \\
0.080 \\
0.083 \\
0.070 \\
0.429 \\
0.326 \\
0.309 \\
0.122 \\
0.196 \\
0.081
\end{bmatrix}
=
\begin{bmatrix}
1 & 1 & 1 & 0 & 1 \\
1 & 1 & 1 & 0 & -1 \\
1 & 1 & 0 & 1 & 1 \\
1 & 1 & 0 & 1 & -1 \\
1 & 1 & -1 & -1 & 1 \\
1 & 1 & -1 & -1 & -1 \\
1 & -1 & 1 & 0 & 1 \\
1 & -1 & 1 & 0 & -1 \\
1 & -1 & 0 & 1 & 1 \\
1 & -1 & 0 & 1 & -1 \\
1 & -1 & -1 & -1 & 1 \\
1 & -1 & -1 & -1 & -1
\end{bmatrix}
\begin{bmatrix}
\beta_0 \\
\beta_1 \\
\beta_2 \\
\beta_3 \\
\beta_4
\end{bmatrix}
$$

This model does not explain the variation in disposition rates, however. The goodness-of-fit statistic X^2_{GOF} is 10.86 with 7 degrees of freedom, which is statistically significant as $\chi^2_{7,0.75} = 9.04$. This means that the model is not portraying the data adequately and we should search for a model that better reflects the observed data, rather than attempting tests concerning individual model parameters.

A possible choice for modifying the model would be to include all the first-order and second-order interaction terms in the model. The interaction parameters using the product approach to modeling and their interpretations are

β_5 = differential effect of prior record × offense 1 interaction

β_6 = differential effect of prior record × offense 2 interaction

β_7 = differential effect of prior record × income interaction

β_8 = differential effect of offense 1 × income interaction

β_9 = differential effect of offense 2 × income interaction

β_{10} = differential effect of prior record × offense 1 × income interaction

β_{11} = differential effect of prior record × offense 2 × income interaction

The first five interaction effects correspond to all possible first-order interactions, and the last two represent second-order interactions.

A *saturated* model $\mathbf{X}_2\boldsymbol{\beta}_2$ containing both main effects and interaction terms is

$$
\begin{bmatrix}
0.433 \\ 0.160 \\ 0.107 \\ 0.080 \\ 0.083 \\ 0.070 \\ 0.429 \\ 0.326 \\ 0.309 \\ 0.122 \\ 0.196 \\ 0.081
\end{bmatrix}
=
\begin{bmatrix}
1 & 1 & 1 & 0 & 1 & 1 & 0 & 1 & 1 & 0 & 1 & 0 \\
1 & 1 & 1 & 0 & -1 & 1 & 0 & -1 & -1 & 0 & -1 & 0 \\
1 & 1 & 0 & 1 & 1 & 0 & 1 & 1 & 0 & 1 & 0 & 1 \\
1 & 1 & 0 & 1 & -1 & 0 & 1 & -1 & 0 & -1 & 0 & -1 \\
1 & 1 & -1 & -1 & 1 & -1 & -1 & 1 & -1 & -1 & -1 & -1 \\
1 & 1 & -1 & -1 & -1 & -1 & -1 & -1 & 1 & 1 & 1 & 1 \\
1 & -1 & 1 & 0 & 1 & -1 & 0 & -1 & 1 & 0 & -1 & 0 \\
1 & -1 & 1 & 0 & -1 & -1 & 0 & 1 & -1 & 0 & 1 & 0 \\
1 & -1 & 0 & 1 & 1 & 0 & -1 & -1 & 0 & 1 & 0 & -1 \\
1 & -1 & 0 & 1 & -1 & 0 & -1 & 1 & 0 & -1 & 0 & 1 \\
1 & -1 & -1 & -1 & 1 & 1 & 1 & -1 & -1 & -1 & 1 & 1 \\
1 & -1 & -1 & -1 & -1 & 1 & 1 & 1 & 1 & 1 & -1 & -1
\end{bmatrix}
\begin{bmatrix}
\beta_0 \\ \beta_1 \\ \beta_2 \\ \beta_3 \\ \beta_4 \\ \beta_5 \\ \beta_6 \\ \beta_7 \\ \beta_8 \\ \beta_9 \\ \beta_{10} \\ \beta_{11}
\end{bmatrix}
$$

The term *saturated* refers to the fact that the number of terms estimated by the model is equal to the degrees of freedom. As a consequence, there is no goodness-of-fit test.

Table 5.2 Summary of Estimated Parameters and Test Statistics for Model $X_2\beta_2$

Parameter	Description	Parameter Estimate	Test Statistic for H_0: $C\beta = 0$	DF
β_0	Constant	0.200	—	—
Main effects				
β_1	Prior arrest record	−0.044	7.52	1
β_2	Offense 1	0.137 ⎫	33.42	2
β_3	Offense 2	−0.045 ⎭		
β_4	Income	0. 60	13.90	1
First-order interaction effects				
β_5	Arrest record × offense 1	0.004 ⎫	0.98	2
β_6	Arrest record × offense 2	−0.017 ⎭		
β_7	Arrest record × income	−0.008	0.22	1
β_8	Offense 1 × income	0.034 ⎫	2.61	2
β_9	Offense 2 × income	−0.006 ⎭		
Second-order interaction effects				
β_{10}	Arrest record × offense 1 × income	0.050 ⎫	3.51	2
β_{11}	Arrest record × offense 2 × income	−0.032 ⎭		

Fitting this model to the data apportions the variation in **F** among all main and interaction effects. In this manner, we may estimate the interactions which are suggested by the lack of fit of the main effects model. Table 5.2 contains the estimates of the parameters and their associated test statistics for model $X_2\beta_2$. None of the interaction terms is statistically significant, although the second-order interaction has a relatively large test statistic associated with it. Based on the results reported in Table 5.2, it is not fruitful to continue with the analysis using this approach. The largest contribution to the sum of squares for lack of fit for the $X_1\beta_1$ model comes from a second-order interaction term. Since deleting this interaction from the model does not achieve a good fit and leaving it in does not achieve parsimony, we will consider another approach: a modular model.

A MODULAR MODEL

Although we have had some suggestion of the presence of interaction in the Charlotte-Mecklenburg County data, we have not asked *why* income might interact with offense charged. Crimes can be characterized by the degree of "seriousness" associated with them. Seriousness is a general concept indicating

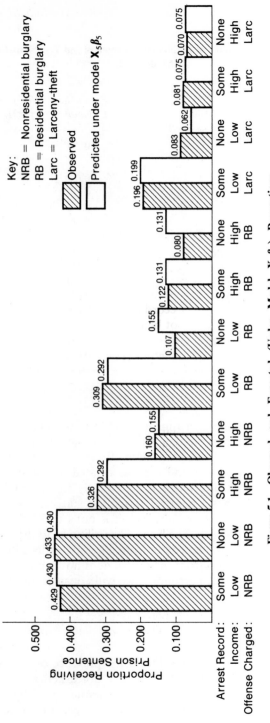

Figure 5.1 Observed and Expected (Under Model $X_5\beta_5$) Proportions Receiving a Prison Sentence by Type of Offense, Defendant's Income, and Prior Arrest Record

the degree of official attention given a particular crime—manifested by the magnitude of the potential fine or prison sentence and by the resources the criminal justice system uses to deal with the crime. Among the offenses studied here, nonresidential burglaries are considered most serious, followed by residential burglaries and then larceny. Clarke and Koch (1976) also note that, other things considered, a property offense committed by a low-income defendant is probably considered more serious because of the possible motives involved. Hence low-income defendants charged with nonresidential burglary probably face more certain penalties because they are perceived as being professional thieves. An alternative interpretation leading to the same result is that low-income defendants cannot provide as good a defense as others and thus incur a higher conviction rate. In either case, a defendant's income level and the offense charged combine to affect disposition.

Figure 5.1 shows this pattern. Nonresidential burglaries among low-income defendants are most likely to receive prison sentences. Prior arrest history does not appear to have a uniform effect on dispositions. For the very serious and less serious crimes, it does not appear to play a role. In the former case, other factors may override a prior arrest record to increase the certainty of conviction. In the latter case, the crime may not be considered serious enough to make a prior record relevant.

An alternative approach to modeling the effects of one or more variables, whose effects are not consistent across the levels of other variables, is to use a *modular* model. This is a model having one or more effects expressed within the levels of one or more other variables.

Table 5.3 defines the parameters of one possible saturated modular model: model $X_3\beta_3$. This model is modular in form because the prior-record-effect parameters, β_7 to β_{12}, depend on their respective income/offense combination.

Table 5.3   Definition of Parameters for Model $X_3\beta_3$

β_1	Mean: low income and nonresidential burglary
β_2	Mean: high income and nonresidential burglary
β_3	Mean: low income and residential burglary
β_4	Mean: high income and residential burglary
β_5	Mean: low income and larceny-theft
β_6	Mean: high income and larceny-theft
β_7	Differential effect of no prior record given β_1
β_8	Differential effect of no prior record given β_2
β_3	Differential effect of no prior record given β_3
β_{10}	Differential effect of no prior record given β_4
β_{11}	Differential effect of no prior record given β_5
β_{12}	Differential effect of no prior record given β_6

Model $\mathbf{X}_3\boldsymbol{\beta}_3$ is

$$
\begin{bmatrix}
0.433 \\
0.160 \\
0.107 \\
0.080 \\
0.083 \\
0.070 \\
0.429 \\
0.326 \\
0.309 \\
0.122 \\
0.196 \\
0.081
\end{bmatrix}
=
\begin{bmatrix}
1 & 0 & 0 & 0 & 0 & 0 & 1 & 0 & 0 & 0 & 0 & 0 \\
0 & 1 & 0 & 0 & 0 & 0 & 0 & 1 & 0 & 0 & 0 & 0 \\
0 & 0 & 1 & 0 & 0 & 0 & 0 & 0 & 1 & 0 & 0 & 0 \\
0 & 0 & 0 & 1 & 0 & 0 & 0 & 0 & 0 & 1 & 0 & 0 \\
0 & 0 & 0 & 0 & 1 & 0 & 0 & 0 & 0 & 0 & 1 & 0 \\
0 & 0 & 0 & 0 & 0 & 1 & 0 & 0 & 0 & 0 & 0 & 1 \\
1 & 0 & 0 & 0 & 0 & 0 & -1 & 0 & 0 & 0 & 0 & 0 \\
0 & 1 & 0 & 0 & 0 & 0 & 0 & -1 & 0 & 0 & 0 & 0 \\
0 & 0 & 1 & 0 & 0 & 0 & 0 & 0 & -1 & 0 & 0 & 0 \\
0 & 0 & 0 & 1 & 0 & 0 & 0 & 0 & 0 & -1 & 0 & 0 \\
0 & 0 & 0 & 0 & 1 & 0 & 0 & 0 & 0 & 0 & -1 & 0 \\
0 & 0 & 0 & 0 & 0 & 1 & 0 & 0 & 0 & 0 & 0 & -1
\end{bmatrix}
\begin{bmatrix}
\beta_1 \\
\beta_2 \\
\beta_3 \\
\beta_4 \\
\beta_5 \\
\beta_6 \\
\beta_7 \\
\beta_8 \\
\beta_9 \\
\beta_{10} \\
\beta_{11} \\
\beta_{12}
\end{bmatrix}
$$

Note that parameters β_1 through β_6 represent the cell mean for income by type of offense combinations and are ordered in terms of severity.

Now we will test hypotheses of the form $\mathbf{C}\boldsymbol{\beta} = \mathbf{0}$ to see if we can summarize the information in Table 5.1. The tests are of two types: (1) tests of the six

**Table 5.4   Summary of Estimated Parameters and Test Statistics
for Model $\mathbf{X}_3\boldsymbol{\beta}_3$**

$\mathbf{C}\boldsymbol{\beta}$	*Description*	b_i *Based on* $\mathbf{X}_3\boldsymbol{\beta}_3$	X^2	*DF*
Arrest effects				
β_7	For NRB low	0.002	0.01	1
β_8	For NRB high	−0.083	2.72	1
β_9	For RB low	−0.101	6.71	1
β_{10}	For RB high	−0.021	0.32	1
β_{11}	For Larc low	−0.056	6.42	1
β_{12}	For Larc high	−0.005	0.09	1
Test of differences in means				
$\beta_1 - \beta_2$	Low − high and NRB		6.35	1
$\beta_2 - \beta_3$	High and NRB − low and RB		0.30	1
$\beta_3 - \beta_4$	Low − high and RB		3.94	1
$\beta_4 - \beta_5$	High and RB − low and Larc		0.79	1
$\beta_5 - \beta_6$	Low − high and Larc		5.08	1
Other tests				
$\beta_8 - \beta_9$	Arrest effect and (NRB and high) − arrest effect and (RB and low)	0.064	0.08	1

specific prior-arrest effects (H_0: $\beta_j = 0, j = 7, 8, \ldots, 12$) and (2) tests of the differences between means (H_0: $\beta_j - \beta_{j+1} = 0, j = 1, 2, \ldots, 5$). There are many other possible tests for the differences in means in the second set. We chose the tests specified here in order to test differences in mean conviction rates based on severity ordering. Table 5.4 summarizes the selected tests.

Table 5.4 shows that the combination of income and offense defines four groups of offenses and that prior-arrest effects are present only for selected crimes. We may summarize the results of Table 5.4 as follows:

- *Group I (most serious):* Nonresidential burglaries by low-income defendants; no prior-arrest effects.
- *Group II:* Nonresidential burglaries by high-income defendants or residential burglaries by low-income defendents; prior-arrest effects present.
- *Group III:* Residential burglaries by high-income defendants or larceny by low-income defendants; prior-arrest effects present.
- *Group IV (least serious):* Larceny by high-income defendants; no prior-arrest effects.

The reduced model is given by $\mathbf{X}_4\boldsymbol{\beta}_4$:

$$
\mathbf{F} = \begin{bmatrix}
1 & 0 & 0 & 0 & 0 & 0 \\
0 & 1 & 0 & 0 & 1 & 0 \\
0 & 1 & 0 & 0 & 1 & 0 \\
0 & 0 & 1 & 0 & 0 & 0 \\
0 & 0 & 1 & 0 & 0 & 1 \\
0 & 0 & 0 & 1 & 0 & 0 \\
1 & 0 & 0 & 0 & 0 & 0 \\
0 & 1 & 0 & 0 & -1 & 0 \\
0 & 1 & 0 & 0 & -1 & 0 \\
0 & 0 & 1 & 0 & 0 & 0 \\
0 & 0 & 1 & 0 & 0 & -1 \\
0 & 0 & 0 & 1 & 0 & 0
\end{bmatrix}
\begin{bmatrix}
\beta_1 \\ \beta_2 \\ \beta_3 \\ \beta_4 \\ \beta_5 \\ \beta_6
\end{bmatrix}
$$

In this model β_1 through β_4 again represent means for income by type of offense combinations (groups I to IV) while β_5 and β_6 represent the differential effects of prior arrest records.

This model provides an adequate fit to the data as the test statistic for the goodness of fit is $X^2_{GOF} = 1.51$, which is less than 7.84 ($\chi^2_{6, 0.75}$). The test of the parameters under model $\mathbf{X}_4\boldsymbol{\beta}_4$ presented in Table 5.5 shows that prior arrest effects β_5 and β_6 explain a significant amount of variation and are similar in magnitude ($\beta_5 = \beta_6$). Furthermore, the four offense and income group means are different from each other.

Table 5.5   Summary of Estimated Parameters and Test Statistics for Model $X_4\beta_4$

Effect	Description	b_i Based on $X_4\beta_4$
β_1	Mean for group I	0.430
β_2	Mean for group II	0.221
β_3	Mean for group III	0.130
β_4	Mean for group IV	0.075
β_5	Prior-arrest effect for group II	−0.094
β_6	Prior-arrest effect for group III	−0.056

Selected Tests

$C\beta$		X^2	DF
β_5	Prior arrest effect for group II	9.25	1
β_6	Prior arrest effect for group IV	6.26	1
$\beta_5 - \beta_6$	Differences in arrest effects	1.01	1
$\beta_1 - \beta_2$	Mean I − mean II	12.18	1
$\beta_2 - \beta_3$	Mean II − mean III	6.38	1
$\beta_3 - \beta_4$	Mean III − mean IV	4.57	1

$X_{\text{GOF}}^2 = 1.51$

Because the prior-arrest effects are statistically equal, we fit a final reduced model $X_5\beta_5$, which contains four group means and a prior-arrest effect:

$$
F = \begin{bmatrix}
1 & 0 & 0 & 0 & 0 \\
0 & 1 & 0 & 0 & 1 \\
0 & 1 & 0 & 0 & 1 \\
0 & 0 & 1 & 0 & 0 \\
0 & 0 & 1 & 0 & 1 \\
0 & 0 & 0 & 1 & 0 \\
1 & 0 & 0 & 0 & 0 \\
0 & 1 & 0 & 0 & -1 \\
0 & 1 & 0 & 0 & -1 \\
0 & 0 & 1 & 0 & 0 \\
0 & 0 & 1 & 0 & -1 \\
0 & 0 & 0 & 1 & 0
\end{bmatrix}
\begin{bmatrix}
\beta_1 \\
\beta_2 \\
\beta_3 \\
\beta_4 \\
\beta_5
\end{bmatrix}
$$

Table 5.6 Summary of Estimated Parameters and Test Statistics for Final Modular Model $X_5\beta_5$

Effect	Description	b_i Based on $X_5\beta_5$	Estimated SE of b_i
β_1	Mean for group I	0.430	0.051
β_2	Mean for group II	0.224	0.031
β_3	Mean for group III	0.131	0.019
β_4	Mean for group IV	0.075	0.017
β_5	Prior arrest effect for groups II and III	−0.069	0.018

Selected Tests

$C\beta$		X^2	DF
β_5	Arrest effect	14.50	1
$\beta_1 - \beta_2$	Mean I − mean II	11.90	1
$\beta_2 - \beta_3$	Mean II − mean III	6.66	1
$\beta_3 - \beta_4$	Mean III − mean IV	4.70	1

$X^2_{GOF} = 2.52$.

The fit of this model is quite good ($X^2_{GOF} = 2.52$), and all parameters contained in the model explain a statistically significant amount of variation (see Table 5.6). Figure 5.1 compares the observed and predicted values based on model $X_5\beta_5$.

COMPARISON OF MODELS

By almost any standard, the final modular model $X_5\beta_5$ provides a better account of the variation in the proportion receiving prison sentences than the main-effects model $X_1\beta_1$. The residual values in most cases are smaller for $X_5\beta_5$ and the goodness-of-fit statistic, which is a function of these errors, is smaller (see Table 5.7). Although both models use 5 degrees of freedom, the goodness-of-fit statistic X^2 for $X_5\beta_5$ is below $\chi^2_{1,0.95}$ whereas the main-effects model $X_1\beta_1$ presents the possibility that a statistically significant parameter has been omitted, because its chi-square value (10.86) exceeds the critical value of $\chi^2_{1,0.95} = 3.84$. Finally, the modular model has substantially smaller standard errors associated with its predicted values than the main effects model.

Although we cannot claim that $X_5\beta_5$ is the true model, we know that the criteria used to evaluate a model—its goodness of fit, the size and pattern of its residuals, and the standard errors of its predicted values—all suggest that $X_5\beta_5$

Table 5.7   Comparison of Main-Effects Model $X_1\beta_1$ with Final Modular Model $X_5\beta_5$

SUBPOPULATION			RESIDUALS		STANDARD ERROR OF $\hat{f}_i$	
Prior Arrest	Offense	Income	$X_1\beta_1$	$X_5\beta_5$	$X_1\beta_1$	$X_5\beta_5$
None	NRB	Low	0.096	0.003	0.407	0.051
None	NRB	High	−0.097	0.005	0.400	0.036
None	RB	Low	−0.057	−0.048	0.324	0.036
None	RB	High	−0.035	−0.051	0.315	0.016
None	Larc	Low	−0.035	0.021	0.227	0.028
None	Larc	High	0.032	−0.004	0.189	0.016
Some	NRB	Low	0.023	−0.002	0.388	0.051
Some	NRB	High	0.002	0.034	0.390	0.035
Some	RB	Low	0.077	0.016	0.310	0.035
Some	RB	High	−0.029	−0.009	0.312	0.016
Some	Larc	Low	0.009	−0.003	0.236	0.029
Some	Larc	High	−0.025	0.006	0.215	0.016

X^2_{GOF} for $X_1\beta_1 = 10.86$.
X^2_{GOF} for $X_5\beta_5 = 2.52$.

is the better choice. The ultimate determination of the choice of model, however, must depend on successive replications of the Charlotte-Mecklenburg County study in which models similar to $X_5\beta_5$ are evaluated in a strict statistical sense rather than being developed from data analysis.

SUMMARY

In this chapter we examined a more complex table based on the data analyzed in Chapter 4, introduced more complex predictor effects of the response function, and raised the issue of whether interaction is present in the data. The lack of fit of a main effects model suggested the presence of interaction.

We subsequently introduced a modular approach to modeling the data that permitted us to estimate prior arrest effects nested within combinations of offense and income. We followed a data-analytic strategy and used the test statistics to identify significant sources of variation and to eliminate unimportant ones. The result of this effort was a 5-degree of freedom modular model $X_5\beta_5$ that adequately characterized the variation in sentencing in a Charlotte-Mecklenburg County, North Carolina, criminal court.

These are the key points covered in the chapter:

- *The problem:* Criminal court dispositions and the relationship to three factor variables.
- *Definition of variables:* Response variable—disposition of the case; factor variables—prior arrest record, type of offense, defendant's income.
- *Interaction:* Main effects model does not fit; complex interaction is present.
- *Modular model:* Interaction effects examined by modeling effects of one factor within levels of other factor variables.
- *Model comparison:* The modular model provides improved fit over the main effects model.

The criminal court data allowed us to focus on the probability that a person was sentenced to prison because there were only two levels for this variable. We did not examine the complement—the probability of not receiving prison sentence—because the model that holds for the probability of receiving a sentence also holds for the probability of not receiving one. In the next chapter we examine a response variable that has more than two levels and consider an alternative function to the proportion.

EXERCISES

5.1.

The Family Nurse Practitioner Study (Wagner and others, 1976) surveyed physicians by means of a mail questionnaire. Nonresponse to mail questionnaires is a problem common to this type of research design and can limit the utility of the study's findings. Table 5.8 summarizes response and nonresponse by physician type and year of graduation. Determine whether response to the questionnaire was selective—that is, dependent on physician type and year of graduation.

    a. Prepare the frequency table in a two-factor, one-response format suitable for input to the GENCAT program. Compute the vector **p** of observed proportions by hand.

    b. Construct an **A** matrix defining the function "proportion of nonresponse" and premultiply **p** to obtain the vector of functions **F**.

    c. Prepare a saturated design matrix containing the following:

1. One general mean term
2. Two physician terms
3. Three year of graduation terms
4. Six physician type × year of graduation interaction terms

Table 5.8 Response to Mail Questionnaire by Year of Graduation and Physician Type

Year of Graduation	INFECTIOUS-DISEASE PEDIATRICIANS		GENERAL PEDIATRICIANS		FAMILY PHYSICIANS	
	Respondents	Nonrespondents	Respondents	Nonrespondents	Respondents	Nonrespondents
Before 1940	5	5	38	33	56	86
1940–1950	9	10	85	58	42	51
1951–1960	23	14	102	60	81	61
After 1960	8	2	49	21	27	35

Source: Kleinbaum and Kupper (1978, p. 481).

Use the GENCAT program to compute the parameter estimates.

d. Construct the following contrast matrices:

1. A 2-DF main effects test for physician type
2. A 3-DF main effects test for year of graduation
3. A 6-DF interaction test
4. An 11-DF test of "total variation" containing the tests (1–3) above

What are the primary sources of variation in response to the mail questionnaire? Will a main effects model fit these data? Explain your answer.

Note: The chi-square statistic for test (4) may be interpreted as a measure of total variation for these data. The chi-square statistics produced by tests (1–3) divided by the chi-square statistic in (4) are approximate measures of the explanatory power of physician type, year of graduation, and interaction. The chi-square statistics produced by tests (1–3) do not sum to test (4) because the variation associated with each source cannot be uniquely partitioned.

e. Fit a reduced model and interpret your results. Write a summary paragraph describing your conclusions regarding the effects of physician type and year of graduation on response to the mail questionnaire.

f. A measure of total variation explained may be constructed for these data by using the following formula:

Proportion of variation explained by reduced model

$$= \frac{X^2 \text{ for "total variation"} - X^2_{GOF} \text{ for reduced model}}{X^2 \text{ for "total variation"}}$$

Note that the chi-square for "total variation" was computed in step (d.4) above. Observe that a saturated model explains 100 percent of the variation in **F** because $X^2_{GOF} = 0$ in this case.

5.2.

Kleinbaum and Kupper (1978) have reported occupational health data from a study of North Carolina textile industry workers (Table 5.9). This health study by Higgins and Koch (1977) reported on the occurrence of byssinosis (also known as "brown lung disease") and its relation to conditions in the workplace, length of employment, and the worker's smoking habits.

a. Analyze these data as a three-factor, one-response problem and construct an **A** matrix that defines the function "positive byssinotic response" for the 12 subpopulations.

Table 5.9   Byssinotic Response by Condition of Workplace, Length of Employment, and Smoking Habits

Workplace	Length of Employment (Years)	Smoking	RESPONSE (BYSSINOTIC OR NOT) Yes	No
Dusty	<10	No	7	119
		Yes	30	203
	10–20	No	3	17
		Yes	16	51
	>20	No	8	64
		Yes	41	110
Not dusty	<10	No	12	1004
		Yes	14	1340
	10–20	No	2	209
		Yes	5	409
	>20	No	8	777
		Yes	19	951

Source: Kleinbaum and Kupper (1978, p. 482).

b. Construct a 12-DF linear model using differential effects coding to estimate the following effects:

Effect	DF	X^2
Mean	1	—
Workplace (W)	1	
Length of employment (E)	2	
Smoking (S)	1	
$W \times E$ interaction	2	
$W \times S$ interaction	1	
$E \times S$	2	
$W \times E \times S$	2	

What is X^2 for the $W \times E \times S$ interaction? Is it statistically significant at the $\alpha = 0.05$ level?

c. Identify the primary sources of variation. What factors are most associated with byssinosis? Do you think that a main effects model would adequately fit these data? Explain your answers.

d. Appendix B describes two alternative methods of coding—differential effects coding and dummy variable coding. Part (b) of this exercise requires a differential effects coding. Reanalyze the byssinosis data by using a saturated model with dummy variable coding. Compare the interpretation of the parameters by using the two methods of coding. Which one is better suited for these data? Explain your answer.

5.3.

Lehnen and Reiss (1978) discuss factors associated with the reporting of personal and household victimizations by respondents to the National Crime Survey (NCS). The investigators analyzed the effects of three factors—(1) the number of prior NCS interviews (the NCS design calls for a total of seven interviews with the same household conducted at 6-month intervals), (2) the number of victimization incidents reported in previous interviews, and (3) the method used in the current interview (telephone versus in-person)—on the number of incidents reported in the present interview. Table 5.10, a three-factor, one-response table, summarizes the findings of the National Crime Survey from 1 July 1972 to 31 December 1975.

Because of the cost of NCS (approximately $6 million in 1976), the Law Enforcement Assistance Administration, the sponsoring agency, is especially interested in substituting telephoning in place of in-person interviewing. The Agency lacks other evaluation criteria, but in past studies it has regarded the method producing the greatest number of incident reports as superior.

Evaluate the relative effect of telephone versus in-person interviewing in light of the known effects of time in sample (previous interviews) and prior incident reporting.

a. Construct an **A** matrix that defines the "proportion reporting one or more incidents in the current interview" for each subpopulation.

b. Develop a saturated design matrix and compute the estimated parameters for the functions created in part (a). Select your coding method so that you may test hypotheses regarding the relative gains (losses) in incident reporting resulting from telephone interviewing once you account for the effects of time in sample and prior incident reporting.

c. Construct suitable **C** matrices to test the total variation in reporting associated with type of interview method, time in sample, and prior incident reporting. Which variables account for the major sources of variation in incident reporting? How does the presence of interaction complicate your interpretation?

Table 5.10 Number of Incidents Reported in National Crime Survey by Type of Interview, Number of Prior Interviews, and Number of Incidents Previously Reported[a]

Type of Interview[b]	Number of Prior Interviews	Number of Previous Incidents	NUMBER OF INCIDENTS REPORTED IN PRESENT INTERVIEW			
			0	*1*	*2*	*3+*
P	0	0	72,427	11,001	2,683	1,249
P	1	0	40,469	3,584	557	139
P	1	1	4,244	764	174	62
P	1	2	782	226	71	25
P	1	3+	246	90	27	23
P	2–3	0	46,250	2,997	384	92
P	2–3	1	7,496	979	188	44
P	2–3	2	1,772	375	97	28
P	2–3	3+	778	225	75	37
P	4+	0	34,365	1,858	189	27
P	4+	1	8,617	890	118	28
P	4+	2	2,655	386	76	17
P	4+	3+	1,484	380	84	42
T	0	0	16,334	1,879	360	126
T	1	0	12,065	713	75	14
T	1	1	1,385	175	29	6
T	1	2	254	45	10	4
T	1	3+	80	22	4	2
T	2–3	0	14,045	628	70	15
T	2–3	1	2,596	234	31	5
T	2–3	2	599	66	18	7
T	2–3	3+	266	51	15	4
T	4+	0	10,127	342	27	8
T	4+	1	2,840	219	21	7
T	4+	2	861	86	13	0
T	4+	3+	599	95	15	5

[a] Includes all nonseries incidents and the last reported incident in a series.
[b] P = person; T = telephone

d. Fit a reduced model and interpret your results. Evaluate the fit of your model and summarize the principal sources of variation. Test for the relative gain (loss) associated with telephone interviewing.

e. What substantive recommendation would you make to the Law Enforcement Assistance Administration regarding the gains or losses in incident reporting expected from telephone interviewing? What limitations would you place on your recommendations?

APPENDIX: GENCAT INPUT

The following listing is for the GENCAT control cards that reproduce the analyses described in this chapter.

```
     5     1     1              CHAPTER 5: PREDICTING SENTENCES
    12     2                    (2F3.0)
   17  13
   21   4
   25   3
   23   2
   77   7
  119   9
   36  27
   31  15
   56  25
   36   5
  119  29
   91   8
     1     2    12     1       1  (2F1.0)
 01
     7     1     5                 (12F2.0)        MAIN EFFECTS MODEL
 010101010101010101010101
 010101010101-1-1-1-1-1-1
 01010000-1-101010000-1-1
 00000101-1-100000101-1-1
 01-101-101-101-101-101-1
     6                         1  REANALYSIS OF DATA
     7     1    12                 (12F2.0)        SATURATED MODEL
 010101010101010101010101
 010101010101-1-1-1-1-1-1
 01010000-1-101010000-1-1
 00000101-1-100000101-1-1
 01-101-101-101-101-101-1
 01010000-1-1-1-100000101
 00000101-1-10000-1-10101
 01-101-101-1-101-101-101
 01-10000-10101-10000-101
 000001-1-101000001-1-101
 01-10000-101-101000001-1
 000001-1-1010000-10101-1
     8     1     1                 (12F1.0)        PRIOR ARREST RECORD
 01
     8     1     2                 (12F1.0)        TOTAL OFFENSE EFFECT
 001
 0001
     8     1     1                 (12F1.0)        INCOME
 00001
     8     1     2                 (12F1.0)        TOTAL ARREST X OFFENSE EFFECT
 000001
 0000001
     8     1     1                 (12F1.0)        ARREST RECORD X INCOME
 00000001
     8     1     2                 (12F1.0)        TOTAL OFFENSE X INCOME EFFECT
 000000001
 0000000001
     8     1     2                 (12F1.0)        TOTAL 3-WAY INTERACTION EFFECT
 00000000001
 000000000001
     6                         1  REANALYSIS OF DATA
     7     1    12                 (12F2.0)        HIERARCHICAL MODEL
```

```
0100000000000100000000000
0001000000000001000000000
0000010000000000010000000
0000000100000000000010000
0000000001000000000000100
0000000000010000000000001
010000000000-10000000000
00010000000000-100000000
0000010000000000-1000000
0000000100000000000-10000
0000000000100000000000-100
000000000001000000000000-1
      8      1      1              (12F2.0)          FOR NRB LOW
000000000000001
      8      1      1              (12F2.0)          FOR NRB HIGH
0000000000000001
      8      1      1              (12F2.0)          FOR RB LOW
000000000000000001
      8      1      1              (12F2.0)          FOR RB HIGH
00000000000000000001
      8      1      1              (12F2.0)          FOR LARC LOW
0000000000000000000001
      8      1      1              (12F2.0)          FOR LARC HIGH
00000000000000000000001
      8      1      1              (12F2.0)          LOW - HIGH : NRB
01-1
      8      1      1              (12F2.0)          HIGH & NRB - LOW & RB
0100-1
      8      1      1              (12F2.0)          LOW - HIGH : RB
000001-1
      8      1      1              (12F2.0)          HIGH & RB - LOW & LARC
00000001-1
      8      1      1              (12F2.0)          LOW - HIGH : LARC
0000000001-1
      8      1      1              (12F2.0)          ARR: NRB & HI - ARR: RB & LOW
00000000000001-1
      6                     1  REANALYSIS OF DATA
      7      1      6              (12F2.0)          REDUCED MODEL
01000000000001
0001010000000000101
0000000101000000000101
000000000001000000000001
00010100000000-1-1
00000000010000000000-1
      8      1      1              (12F2.0)          ARREST EFFECT
0000000001
      8      1      1              (12F2.0)          MEAN I - MEAN II
01-1   .
      8      1      1              (12F2.0)          MEAN II - MEAN III
0001-1
      8      1      1              (12F2.0)          MEAN III - MEAN IV
000001-1
```

6

Mean Scores

The preceding chapter treated the situation in which there are two possible outcomes for the response variable. In this chapter we consider an extension in which the response variable is a discrete variable with more than two outcomes and the categories are ordered. We will take advantage of the ordering to create a single score for each subpopulation and then analyze the scores with the same methods used in Chapters 4 and 5.

MEASURING HMO USAGE

In 1973, Congress passed Public Law 93-222, which was intended to foster the development of health maintenance organizations (HMOs). Health maintenance organizations differ from the traditional health care delivery system in that (1) they are comprehensive and (2) their financing is based on a fixed fee per person. An HMO makes a contract with a person that obligates it to provide all the health services a person requires in return for a prepaid fee. This method differs from the traditional system, which is based on fee for service and requires a patient to interact with a number of different providers in order to obtain comprehensive health care.

As a result of the Health Maintenance Organization Act of 1973, interest in the development of HMOs began to increase. Those starting an HMO face a variety of planning questions:

1. What are the number and distribution of physicians?
2. What are the number and type of nurses and other health workers?
3. What is the number of clinics, and how will they be equipped?
4. What are the number and distribution of hospital beds?

Since HMOs are responsible for delivering comprehensive care to their members, they must know the characteristics of the group to which they will be marketing their plan. The needs of the elderly, for example, differ from those of a population consisting mostly of young families. Data from existing HMOs can provide evidence on these different needs and suggest answers to the four questions raised above.

The following discussion suggests one way that existing data can be used to estimate demand for services. The data are from the Portland, Oregon, Region of the Kaiser Foundation Health Plan. This HMO prototype has collected data on the use experience—that is, the demand for services—of a 5 percent sample of its subscriber units since September 1966 (Greenlick and others, 1968). The data include both the inpatient and outpatient components of the system, and these contacts can be further divided by type, time, date, place of service, and who provided the services. Apart from the use experience of its members, Kaiser records demographic and socioeconomic data.

A person planning to hire personnel for an HMO could use the Kaiser data to estimate the staff required to service the target population and the use expected from different subgroups. Overall use can be examined as well as any of the component types such as obstetrics-gynecology, pediatrics, and orthopedics. Three demographic variables of special interest are age and sex of the head of household and family size—data that are readily available to health planners. Additional variables of possible importance are income and education; nonresponse, however, is a problem limiting the utility of these variables.

The data in Table 6.1 represent the number of outpatient contacts with pediatric physicians made in 1970 by families in the Kaiser 5 percent sample who were enrolled for 12 months during 1970. They are categorized by age and sex of the head of household and by the number of people in the family unit enrolled in the Kaiser Plan. There are four levels for the age variable, two levels for the sex variable, and five levels for the family size variable. Considering these three variables simultaneously would yield 40 ($4 \times 2 \times 5$) subgroups of the population. A preliminary examination of the three-way cross-tabulation of the data reveals that few families with seven or more members were enrolled in the plan. Only one subgroup involving seven or more family members enrolled in the plan contains more than 25 observations—the minimum subgroup size for inclusion in a study (as suggested in Chapter 2). Therefore we

Table 6.1 Number of Outpatient Contacts with Pediatric Physicians Made by Families During 1970 (by Subscriber's Age and Sex and Number of Members Enrolled in Plan)

Age of Subscriber	Sex of Subscriber	Number of Family Members Enrolled	NUMBER OF PEDIATRIC CONTACTS							Total
			0	1	2	3	4–6	7–9	≥10	
<25	M	1	72	1	0	0	1	1	0	75
		2	27	2	2	1	0	1	1	34
		3	2	5	1	3	5	0	9	25
		4–6	7	2	0	0	5	3	14	31
		≥7	0	0	0	0	1	0	0	1
	F	1	90	0	3	2	0	0	1	96
		2	19	2	1	3	1	4	6	36
		3	2	1	1	2	7	8	9	30
		4–6	3	1	3	1	3	3	12	26
		≥7	1	1	1	0	0	0	0	3
25–44	M	1	87	0	0	0	0	0	0	87
		2	41	1	0	0	0	0	0	42
		3	20	3	4	6	10	5	22	70
		4–6	30	25	14	24	49	33	135	310
		≥7	0	0	1	2	3	5	29	40
	F	1	67	0	0	0	0	0	0	67
		2	32	1	1	2	2	1	0	39
		3	20	0	4	2	11	3	12	52
		4–6	17	9	7	15	16	13	46	123
		≥7	0	0	1	1	1	2	12	17
45–59	M	1	68	0	0	0	0	0	0	68
		2	125	0	0	0	0	0	0	125
		3	34	5	3	6	4	3	2	57
		4–6	34	8	13	4	10	7	11	87
		≥7	3	1	0	1	4	2	4	15
	F	1	114	0	0	0	0	0	0	114
		2	76	1	1	0	1	0	2	81
		3	23	2	1	1	1	0	3	31
		4–6	6	2	3	2	5	2	3	23
		≥7	1	0	0	0	0	1	0	2
≥60	M	1	88	0	0	0	0	0	0	88
		2	186	0	0	0	0	0	0	186
		3	5	0	0	0	0	0	0	5
		4–6	1	0	0	0	2	0	0	3
		≥7	0	0	0	0	0	0	0	0
	F	1	208	0	0	0	0	0	0	208
		2	31	0	1	0	0	0	0	32
		3	1	0	0	0	0	0	0	1
		4–6	0	0	0	0	0	0	1	1
		≥7	0	0	0	0	0	0	0	0

could exclude these groups or, if they appear similar to the groups of four to six family members, we could combine them into a group of four or more family members. Inspection of the data suggests that the data in the subgroups of four to six members and seven or more members show the same trend, and therefore we combine the two groups into the subgroup of four or more family members.

Moreover, 12 subgroups had extremely low levels of pediatric contacts. These groups consisted of families with subscribers 60 years old or older, families with a single member enrolled, or families with a male subscriber and a total of two members enrolled. These families are unlikely to have young children and hence would have little need for a pediatrician—as shown by their accounting for only 1.5 percent of the total pediatric contacts. Since our goal is to develop a model that tells HMO planners how many pediatricians are needed, these families can be excluded from the analysis without a severe bias in the results. As a consequence of these decisions, 1133 of the original 2331 families remain to be analyzed (Table 6.2).

Table 6.2   Data from Table 6.1 with Subpopulations Having Few Observations Either Combined or Eliminated

Age of Subscriber	Sex of Subscriber	Number of Family Members Enrolled	NUMBER OF PEDIATRIC CONTACTS						
			0	1	2	3	4–6	7–9	≥10
<25	M	2	27	2	2	1	0	1	1
		3	2	5	1	3	5	0	9
		≥4	7	2	0	0	6	3	14
	F	2	19	2	1	3	1	4	6
		3	2	1	1	2	7	8	9
		≥4	4	2	4	1	3	3	12
25–44	M	3	20	3	4	6	10	5	22
		≥4	30	25	15	26	52	38	164
	F	2	32	1	1	2	2	1	0
		3	20	0	4	2	11	3	12
		≥4	17	9	8	16	17	15	58
45–59	M	3	34	5	3	6	4	3	2
		≥4	37	9	13	5	14	9	15
	F	2	76	1	1	0	1	0	2
		3	23	2	1	1	1	0	3
		≥4	7	2	3	2	5	3	3

FORMATION OF MEANS

We begin the analysis by reducing the seven response categories to a single measure. Our method of reduction takes advantage of the ordering of the levels of the response variables. This process consists of forming a mean score for each of the subgroups considered in this analysis. After forming the mean for each subgroup, we analyze the means in a fashion similar to the analysis of variance procedure used with a continuous dependent variable.

To form the mean for each population subgroup, we focus on the table of observed probabilities (Table 6.3) that correspond to the frequency counts. Each level of the number of contacts variable is assigned a score, which is its value or the value of the midpoint of its interval. For the last interval, which is open-ended, we have assigned the value of 20, the approximate mean number of contacts for this interval. To form the mean for each population subgroup, weight the score assigned each level of the response variable by the proportion

Table 6.3   Observed Probabilities of Contacts by Age, Sex, and Number of Family Members Enrolled in Plan

Age of Subscriber	Sex of Subscriber	Number of Family Members Enrolled	NUMBER OF PEDIATRIC CONTACTS						
			0	1	2	3	4–6	7–9	≥10
<25	M	2	0.79	0.06	0.06	0.03	0.00	0.03	0.03
		3	0.08	0.20	0.04	0.12	0.20	0.00	0.36
		≥4	0.22	0.06	0.00	0.00	0.19	0.09	0.44
	F	2	0.53	0.06	0.03	0.08	0.03	0.11	0.17
		3	0.07	0.03	0.03	0.07	0.23	0.27	0.30
		≥4	0.14	0.07	0.14	0.03	0.10	0.10	0.41
25–44	M	3	0.29	0.04	0.06	0.09	0.14	0.07	0.31
		≥4	0.09	0.07	0.04	0.07	0.15	0.11	0.47
	F	2	0.82	0.03	0.03	0.05	0.05	0.03	0.00
		3	0.38	0.00	0.08	0.04	0.21	0.06	0.23
		≥4	0.12	0.06	0.06	0.11	0.12	0.11	0.41
45–59	M	3	0.60	0.09	0.05	0.11	0.07	0.05	0.04
		≥4	0.36	0.09	0.13	0.05	0.14	0.09	0.15
	F	2	0.94	0.01	0.01	0.00	0.01	0.00	0.02
		3	0.74	0.06	0.03	0.03	0.03	0.00	0.10
		≥4	0.28	0.08	0.12	0.08	0.20	0.12	0.12

of the families exhibiting that response. In mathematical notation, we have

$$\text{Mean for subgroup } i = f_i = \sum_{j=1}^{7} p_{ij}s_j$$

where s_j is the score for the jth response level. The s_j's are 0, 1, 2, 3, 5, 8, and 20, respectively. The mean for population subgroup 1 is

$$
\begin{aligned}
f_1 = \sum_{j=1}^{7} p_{1j}s_j &= (p_{11} \times 0) + (p_{12} \times 1) + (p_{13} \times 2) + (p_{14} \times 3) \\
&\quad + (p_{15} \times 5) + (p_{16} \times 8) + (p_{17} \times 20) \\
&= (0.79 \times 0) + (0.06 \times 1) + (0.06 \times 2) + (0.03 \times 3) \\
&\quad + (0.00 \times 5) + (0.03 \times 8) + (0.03 \times 20) \\
&= 1.09
\end{aligned}
$$

Each score is weighted by the proportion of the subgroup in that response category and then summed to provide the mean number of contacts for the

Table 6.4   Observed, Predicted, and Residual Means for Data in Table 6.2

Subgroup		Number in Subgroup	Observed	Predicted	Residuals
<25, M,	2	34	1.09	2.07	−0.98
	3	25	8.84	6.68	2.16
	≥4	32	10.50	10.92	−0.42
F,	2	36	4.72	2.78	1.94
	3	30	9.60	7.38	2.22
	≥4	29	10.07	11.63	−1.56
25–44, M,	3	70	7.99	5.44	2.55
	≥4	350	11.36	9.68	1.68
F,	2	39	0.69	1.54	−0.85
	3	52	6.40	6.15	0.26
	≥4	140	10.27	10.39	−0.12
45–59, M,	3	57	1.98	3.00	−1.02
	≥4	102	4.82	7.24	−2.42
F,	2	81	0.59	−0.89	1.49
	3	31	2.32	3.71	−1.39
	≥4	25	4.92	7.95	−3.03

subgroup. The "observed" column of Table 6.4 contains the mean for each subgroup.

THE PREDICTION MODEL

We now wish to examine the data in Table 6.4 to determine the relationship between the mean number of contacts and the selected demographic variables. We begin with a simple linear relationship:

$$\text{Mean score} = \text{constant} + \text{an age effect} + \text{a sex effect} + \text{a size of membership effect}$$

Using the linear model framework $\mathbf{F} = \mathbf{X}_1\boldsymbol{\beta}_1$, we have $\mathbf{F}$, the vector of means, and $\boldsymbol{\beta}_1$, the vector of the constant and differential effects, where

$$\boldsymbol{\beta}_1 = \begin{bmatrix} \beta_0 \\ \beta_1 \\ \beta_2 \\ \beta_3 \\ \beta_4 \\ \beta_5 \end{bmatrix} = \begin{bmatrix} \text{constant} \\ \text{differential effect of ages} <25 \\ \text{differential effect of ages } 25\text{--}44 \\ \text{differential effect of males} \\ \text{differential effect of membership size 2} \\ \text{differential effect of membership size 3} \end{bmatrix}$$

The design matrix $\mathbf{X}_1$ and the vector of effects $\boldsymbol{\beta}_1$ are

$$\underset{(16 \times 6)(6 \times 1)}{\mathbf{X}_1 \ \boldsymbol{\beta}_1} = \begin{bmatrix} 1 & 1 & 0 & 1 & 1 & 0 \\ 1 & 1 & 0 & 1 & 0 & 1 \\ 1 & 1 & 0 & 1 & -1 & -1 \\ 1 & 1 & 0 & -1 & 1 & 0 \\ 1 & 1 & 0 & -1 & 0 & 1 \\ 1 & 1 & 0 & -1 & -1 & -1 \\ 1 & 0 & 1 & 1 & 0 & 1 \\ 1 & 0 & 1 & 1 & -1 & -1 \\ 1 & 0 & 1 & -1 & 1 & 0 \\ 1 & 0 & 1 & -1 & 0 & 1 \\ 1 & 0 & 1 & -1 & -1 & -1 \\ 1 & -1 & -1 & 1 & 0 & 1 \\ 1 & -1 & -1 & 1 & -1 & -1 \\ 1 & -1 & -1 & -1 & 1 & 0 \\ 1 & -1 & -1 & -1 & 0 & 1 \\ 1 & -1 & -1 & -1 & -1 & -1 \end{bmatrix} \begin{bmatrix} \beta_0 \\ \beta_1 \\ \beta_2 \\ \beta_3 \\ \beta_4 \\ \beta_5 \end{bmatrix}$$

The analysis consists of finding the estimate of $\boldsymbol{\beta}_1$ and determining whether this estimate results in good agreement between the observed means and the predicted means (based on $\mathbf{X}_1\mathbf{b}_1$). If the model fits—that is, if there is good agreement between the observed and predicted means—we will perform additional tests to determine whether all three independent variables or factors are required in the specification of the model.

The prediction equation resulting from using the age, sex, and family size variables is

$$\hat{f}_i = 5.27 + \begin{Bmatrix} 1.64 \text{ if } <25 \\ 0.40 \text{ if } 25\text{--}44 \\ -2.04 \text{ if } 45\text{--}59 \end{Bmatrix} + \begin{Bmatrix} -0.35 \text{ if male} \\ 0.35 \text{ if female} \end{Bmatrix}$$

$$+ \begin{Bmatrix} -4.48 \text{ if 2 members are enrolled} \\ 0.12 \text{ if 3 members are enrolled} \\ 4.36 \text{ if 4 or more members are enrolled} \end{Bmatrix}$$

For the sixth subgroup, which consists of female subscribers less than 25 years old in families with four or more members enrolled, the predicted mean is

$$b_0 + b_1 - b_3 - b_4 - b_5 = 5.27 + 1.64 + 0.35 - (-4.48) - (0.12)$$
$$\doteq 11.63$$

Table 6.4 contains the observed, predicted, and residual means for the 16 subpopulations. Note that the agreement between observed and predicted means is not very good. This observation is supported by the goodness-of-fit test statistic $X^2_{\text{GOF}} = 83.10$. If the model is true and the observed discrepancies are due to random variation, this statistic is asymptotically distributed as a chi-square statistic with 10 degrees of freedom. Since $X^2_{\text{GOF}} = 83.10$ is larger than $\chi^2_{10, 0.75} = 12.55$ (the seventy-fifth percentile of the chi-square distribution with 10 degrees of freedom), we reject the model $\mathbf{X}_1\boldsymbol{\beta}_1$.

The next step is to see if we can improve the fit by developing an alternative model. Since age and sex variables often interact, we have reason to add this interaction to the model. Because other interactions possible in this example— age × family size, sex × family size, and age × sex × family size—are not reported consistently in the literature, we have no strong theoretical reason for including them. Moreover, there are no patterns observed in the residuals that would suggest other possible factors to be included. Therefore our next model includes the age and sex interaction but no others.

The new design matrix $\mathbf{X}_2$ and the new $\boldsymbol{\beta}_2$ are

$$
\underset{(16 \times 8)(8 \times 1)}{\mathbf{X}_2 \quad \boldsymbol{\beta}_2} =
\begin{bmatrix}
1 & 1 & 0 & 1 & 1 & 0 & 1 & 0 \\
1 & 1 & 0 & 1 & 0 & 1 & 1 & 0 \\
1 & 1 & 0 & 1 & -1 & -1 & 1 & 0 \\
1 & 1 & 0 & -1 & 1 & 0 & -1 & 0 \\
1 & 1 & 0 & -1 & 0 & 1 & -1 & 0 \\
1 & 1 & 0 & -1 & -1 & -1 & -1 & 0 \\
1 & 0 & 1 & 1 & 0 & 1 & 0 & 1 \\
1 & 0 & 1 & 1 & -1 & -1 & 0 & 1 \\
1 & 0 & 1 & -1 & 1 & 0 & 0 & -1 \\
1 & 0 & 1 & -1 & 0 & 1 & 0 & -1 \\
1 & 0 & 1 & -1 & -1 & -1 & 0 & -1 \\
1 & -1 & -1 & 1 & 0 & 1 & -1 & -1 \\
1 & -1 & -1 & 1 & -1 & -1 & -1 & -1 \\
1 & -1 & -1 & -1 & 1 & 0 & 1 & 1 \\
1 & -1 & -1 & -1 & 0 & 1 & 1 & 1 \\
1 & -1 & -1 & -1 & -1 & -1 & 1 & 1
\end{bmatrix}
\begin{bmatrix}
\beta_0 \\ \beta_1 \\ \beta_2 \\ \beta_3 \\ \beta_4 \\ \beta_5 \\ \beta_6 \\ \beta_7
\end{bmatrix}
$$

where β_0 to β_5 are identical to their definitions in $\boldsymbol{\beta}_1$, and β_6 and β_7 are the differential interaction effects of age and sex measured from their mean.

The test statistic for this model is $X^2_{\text{GOF}} = 24.67$ and the critical value for the goodness-of-fit test is $\chi^2_{8,0.75} = 10.22$. Since 24.67 is greater than 10.22, we again reject the model. Note that this model is a big improvement over the previous one in the sense that X^2_{GOF} is smaller, but it still does not provide a good fit to the data. We might try additional models with other terms added, but lacking substantive guidance we could easily undertake many fruitless analyses producing vague, if not misleading, conclusions.

An alternative approach is to conceptualize the problem. We can see that the age of subscriber variable and family size variables are probably not ideal measures of conditions that determine pediatric visits. Variables such as age of youngest child and number of children enrolled would undoubtedly characterize the predictive conditions better. We therefore restructure the problem to include age of subscriber, sex of subscriber, and two new variables—number of children and age of youngest child. The number of family members enrolled is excluded. Table 6.5 summarizes the new data set. In the next section we analyze a new model that incorporates this new combination of variables.

Table 6.5 Number of Pediatric Physican Contacts by Subscriber's Age and Sex, Number of Children Enrolled, and Age of Youngest Child

Age of Subscriber	Sex of Subscriber	Number of Children	Age of Youngest	NUMBER OF PEDIATRIC CONTACTS						
				0	1	2	3	4–6	7–9	≥10
<25	M	1	<5	2	7	1	2	4	1	9
		>1	<5	3	0	0	0	6	2	11
	F	1	<5	2	1	1	3	6	7	10
		>1	<5	0	2	2	1	2	6	14
25–44	M	1	<5	1	2	2	6	8	4	18
		>1	<5	8	5	6	9	24	30	131
			5–9	9	16	8	11	26	5	28
			>9	16	4	2	6	3	4	6
	F	1	<5	2	0	2	1	6	3	8
		>1	<5	9	2	3	9	13	11	41
			5–9	8	6	4	8	7	4	18
			>9	9	1	2	1	2	1	3
45–59	M	1	>9	31	3	3	4	3	3	1
		>1	5–9	5	5	7	3	8	1	4
			>9	32	4	6	2	6	5	3
	F	1	>9	32	2	2	1	1	0	2

DEFINING NEW FACTOR VARIABLES

Because of the previous analyses and the relatively small cell sample sizes within subgroups that occur when four variables are used, we chose only two levels for the number of children variable: one child and more than one child. The age of the youngest child is grouped into three levels: less than 5, 5 to 9, and greater than 9 years. We deleted a number of subgroups due to small sample sizes, attempted to retain about 20 cases in each subpopulation, and deleted subgroups that had almost no contacts. There were 803 families of size 1, for example, that had a total of 46 pediatric contacts. We deleted families in which the subscriber was 60 or older and families with no children. The 16 subgroups used in this analysis account for 91 percent of the pediatric contacts.

We begin by considering the simple model containing only the main effects. The design matrix and vector of parameters are

$$
\mathbf{X}_3\boldsymbol{\beta}_3 =
\begin{bmatrix}
1 & 1 & 0 & 1 & 1 & 1 & 0 \\
1 & 1 & 0 & 1 & -1 & 1 & 0 \\
1 & 1 & 0 & -1 & 1 & 1 & 0 \\
1 & 1 & 0 & -1 & -1 & 1 & 0 \\
1 & 0 & 1 & 1 & 1 & 1 & 0 \\
1 & 0 & 1 & 1 & -1 & 1 & 0 \\
1 & 0 & 1 & 1 & -1 & 0 & 1 \\
1 & 0 & 1 & 1 & -1 & -1 & -1 \\
1 & 0 & 1 & -1 & 1 & 1 & 0 \\
1 & 0 & 1 & -1 & -1 & 1 & 0 \\
1 & 0 & 1 & -1 & -1 & 0 & 1 \\
1 & 0 & 1 & -1 & -1 & -1 & -1 \\
1 & -1 & -1 & 1 & 1 & -1 & -1 \\
1 & -1 & -1 & 1 & -1 & 0 & 1 \\
1 & -1 & -1 & 1 & -1 & -1 & -1 \\
1 & -1 & -1 & -1 & 1 & -1 & -1
\end{bmatrix}
\begin{bmatrix}
\beta_0 \\ \beta_1 \\ \beta_2 \\ \beta_3 \\ \beta_4 \\ \beta_5 \\ \beta_6
\end{bmatrix}
$$

$$
\boldsymbol{\beta}_3 =
\begin{bmatrix}
\text{constant} \\
\text{differential effect of age} <25 \\
\text{differential effect of age } 25\text{--}44 \\
\text{differential effect of male} \\
\text{differential effect of one child} \\
\text{differential effect of youngest child's age} <5 \\
\text{differential effect of youngest child's age } 5\text{--}9
\end{bmatrix}
$$

The statistic for testing the fit of this model has the value of 10.97 with 9 degrees of freedom. Since this value is less than $\chi^2_{9,\,0.75} = 11.39$, we have evidence of an acceptable model. There is some lack of fit, but not enough to keep us from testing hypotheses about model effects. Table 6.6 contains the observed, predicted, and residual values. There does not appear to be any pattern to the residuals—additional confirmation that the model is adequate. The parameter estimates that generate this model and their estimated standard errors are

$$
\mathbf{b}_3 = \begin{bmatrix} 6.58 \\ -0.17 \\ 1.27 \\ 0.27 \\ -0.76 \\ 4.56 \\ -1.03 \end{bmatrix} \qquad \mathrm{SE}(\mathbf{b}_3) = \begin{bmatrix} 0.35 \\ 0.65 \\ 0.41 \\ 0.26 \\ 0.28 \\ 0.45 \\ 0.42 \end{bmatrix}
$$

Table 6.6   Observed, Predicted, and Residual Estimates Based on Model $X_3\beta_3$

Subgroup		Number in Subgroup	Observed	Predicted	Residuals
<25, M,	1, <5	26	8.58	10.47	−1.89
	>1, <5	22	12.09	12.00	0.10
F,	1, <5	30	9.93	9.92	0.01
	>1, <5	27	12.85	11.45	1.40
25–44, M,	1, <5	41	11.12	11.91	−0.79
	>1, <5	213	14.20	13.44	0.76
	5–9	103	7.72	7.85	−0.13
	>9	41	4.71	5.36	−0.66
F,	1, <5	22	10.05	11.14	−1.32
	>1, <5	88	11.46	12.89	−1.44
	5–9	55	8.45	7.30	1.15
	>9	19	4.53	4.82	−0.29
45–59, M,	1, >9	48	1.67	1.47	0.20
	>1,5–9	33	4.73	5.48	−0.76
	>9	58	2.62	3.00	−0.38
F,	1, >9	40	1.35	0.92	0.43

The prediction equation is

$$\text{Predicted mean} = 6.58 + \begin{Bmatrix} -0.17 \text{ if } <25 \\ 1.27 \text{ if } 25-44 \\ -1.10 \text{ if } 45-59 \end{Bmatrix} + \begin{Bmatrix} 0.27 \text{ if male} \\ -0.27 \text{ if female} \end{Bmatrix}$$
$$+ \begin{Bmatrix} -0.76 \text{ if } 1 \text{ child} \\ 0.76 \text{ if } >1 \text{ child} \end{Bmatrix} + \begin{Bmatrix} 4.56 \text{ if youngest } <5 \\ -1.03 \text{ if youngest } 5-9 \\ -3.53 \text{ if youngest } >9 \end{Bmatrix} \quad (6.1)$$

Given these four characteristics of families, we can predict reasonably well the number of pediatric physician contacts that a family will have in a year. The "predicted" column of Table 6.6 presents these estimates for each subpopulation.

Since the model fits the data in Table 6.5, we may test whether all four factors are required in the prediction process and test the four hypotheses

H_{01}: no difference exists among the age of subscriber effects

H_{02}: no difference exists between the sex of subscriber effects

H_{03}: no difference exists between the number of children effects

H_{04}: no difference exists among the age of the youngest child effects

These four hypotheses can be written as

$$H_{01}: \quad \beta_1 = 0 \qquad H_{02}: \quad \beta_3 = 0$$
$$\beta_2 = 0$$
$$H_{03}: \quad \beta_4 = 0 \qquad H_{04}: \quad \beta_5 = 0$$
$$\beta_6 = 0$$

The appropriate **C** matrices for testing hypotheses are

$$\mathbf{C}_1 = \begin{bmatrix} 0 & 1 & 0 & 0 & 0 & 0 & 0 \\ 0 & 0 & 1 & 0 & 0 & 0 & 0 \end{bmatrix}$$
$$\mathbf{C}_2 = \begin{bmatrix} 0 & 0 & 0 & 1 & 0 & 0 & 0 \end{bmatrix}$$
$$\mathbf{C}_3 = \begin{bmatrix} 0 & 0 & 0 & 0 & 1 & 0 & 0 \end{bmatrix}$$
$$\mathbf{C}_4 = \begin{bmatrix} 0 & 0 & 0 & 0 & 0 & 1 & 0 \\ 0 & 0 & 0 & 0 & 0 & 0 & 1 \end{bmatrix}$$

Table 6.7   Test Statistics for Model $X_3\beta_3$

Hypothesis	Test Statistic X^2	DF
No age of subscriber effect	10.91	2
No sex of subscriber effect	1.15	1
No number of children effect	7.20	1
No age of youngest child effect	105.76	2

The four test statistics (Table 6.7) indicate that the sex of the subscriber could be eliminated from the equation without a great loss of information. Since the other test statistics are all larger than their corresponding ninety-ninth percentile of the chi-square distribution, we should retain them in the model.

A REDUCED MODEL

We can do a further analysis with the sex of the subscriber deleted from consideration. By pooling the males and females, there would now be 9 subgroups instead of the original 16. Table 6.8 presents these pooled data. Pooling has one advantage—it permits us to retain more families for analysis than reported

Table 6.8   Number of Pediatric Physician Contacts by Subscriber's Age, Number of Children Enrolled, and Age of Youngest Child

Age of Subscriber	Number of Children Enrolled	Age of Youngest	NUMBER OF PEDIATRIC CONTACTS						
			0	1	2	3	4–6	7–9	≥10
25	1	<5	4	8	2	5	10	8	19
	>1	<5	3	2	2	1	8	8	25
25–44	1	<5	3	2	4	7	14	7	26
	>1	<5	17	7	9	18	37	41	172
		5–9	17	22	12	19	33	9	46
		>9	25	5	4	7	5	5	9
45–59	1	>9	63	5	5	5	4	3	3
	>1	5–9	7	5	8	4	10	3	6
		>9	40	6	8	4	9	6	4

in Table 6.6. Note that in Table 6.8 we count all the people in each combination of the three factors instead of simply summing the people in Table 6.6.

The model under consideration, $X_4\beta_4$, is the same as in the previous section except for the deletion of the sex variable. The matrices X_4 and β_4 that represent this model are

$$X_4\beta_4 = \begin{bmatrix} 1 & 1 & 0 & 1 & 1 & 0 \\ 1 & 1 & 0 & -1 & 1 & 0 \\ 1 & 0 & 1 & 1 & 1 & 0 \\ 1 & 0 & 1 & -1 & 1 & 0 \\ 1 & 0 & 1 & -1 & 0 & 1 \\ 1 & 0 & 1 & -1 & -1 & -1 \\ 1 & -1 & -1 & 1 & -1 & -1 \\ 1 & -1 & -1 & -1 & 0 & 1 \\ 1 & -1 & -1 & -1 & -1 & -1 \end{bmatrix} \begin{bmatrix} \beta_0 \\ \beta_1 \\ \beta_2 \\ \beta_3 \\ \beta_4 \\ \beta_5 \end{bmatrix}$$

where

$$\begin{bmatrix} \beta_0 \\ \beta_1 \\ \beta_2 \\ \beta_3 \\ \beta_4 \\ \beta_5 \end{bmatrix} = \begin{bmatrix} \text{constant} \\ \text{differential effect of age} <25 \\ \text{differential effect of age } 25\text{--}44 \\ \text{differential effect of one child} \\ \text{differential effect of youngest} <5 \\ \text{differential effect of youngest } 5\text{--}9 \end{bmatrix}$$

The test statistic for measuring the fit of this model is $X^2_{GOF} = 2.81$, which is less than $\chi^2_{3,0.75} = 4.11$. The agreement of the observed and predicted means is good (Table 6.9).

The parameter estimates and their estimated standard errors are

$$\mathbf{b}_4 = \begin{bmatrix} 6.64 \\ -0.33 \\ 1.20 \\ -0.84 \\ 4.55 \\ -0.90 \end{bmatrix} \qquad \text{SE}(\mathbf{b}_4) = \begin{bmatrix} 0.34 \\ 0.64 \\ 0.40 \\ 0.27 \\ 0.45 \\ 0.41 \end{bmatrix}$$

These values are very close to the estimates of the previous section—which reinforces the notion that the sex variable does not add much information.

Table 6.9   Observed, Predicted, and Residuals Based on Model $X_4\beta_4$

	SUBGROUP		MEANS			Standard
						Error of
Age of	Number of	Age of				
Subscriber	Children	Youngest	Observed	Predicted	Residuals	Predicted
<25	1	<5	9.30	10.02	−0.72	0.82
	>1	<5	12.51	11.71	0.80	0.82
25–44	1	<5	10.75	11.54	−0.80	0.60
	>1	<5	13.40	13.23	0.16	0.42
		5–9	7.97	7.78	0.19	0.58
		>9	4.65	5.03	−0.37	0.72
45–59	1	>9	1.52	1.27	0.25	0.39
	>1	5–9	5.28	5.72	−0.44	0.75
		>9	2.69	2.96	−0.27	0.45

The prediction equation is now

$$\text{Predicted mean} = 6.64 + \begin{cases} -0.33 \text{ if age } <25 \\ 1.20 \text{ if age } 25\text{--}44 \\ -0.87 \text{ if age } 45\text{--}59 \end{cases}$$

$$+ \begin{cases} -0.84 \text{ if one child} \\ 0.84 \text{ if more than one child} \end{cases}$$

$$+ \begin{cases} 4.55 \text{ if youngest } <5 \\ -0.90 \text{ if youngest } 5\text{--}9 \\ -3.65 \text{ if youngest } >9 \end{cases}$$

Table 6.10 shows the results of testing whether all three factors are required.

Table 6.10   Test Statistics for Selected Hypotheses Under Model $X_4\beta_4$

| | Test Statistic | |
Hypothesis	X^2	DF
No age effect	9.80	2
No number of children effect	10.15	1
No age of youngest child effect	107.05	2

These statistics indicate that all three factors are in fact required in the specification of the model.

USING THE MODEL

We have identified three variables that suitably predict demand for pediatric services. These variables are commonly available, and an HMO planner may therefore estimate the expected demand and thereby determine how many staff are required. Calculate the predicted pediatrician use as follows:

1. Group the projected population of families into the categories considered in the analysis.
2. Apply the prediction equation for each cell (combination of the three factors) to determine the predicted use.
3. Sum the predicted use across the cells.
4. Inflate the predicted use by dividing the total predicted use by 0.909 to account for the contacts associated with the groups not considered.
5. Divide the predicted total number of contacts by the number of contacts a pediatrician can reasonably deal with in a year.

Consider, for example, just three population subgroups

Age	Number of Children	Age of Youngest	Number of Families	Predicted Use and SD per Family per Year
25–44	1	<5	1000	11.54 ± 0.60 contacts
25–44	>1	<5	1500	13.23 ± 0.42 contacts
25–44	>1	5–9	3000	7.78 ± 0.58 contacts

The predicted total use for these three subgroups is

$$(1000 \times 11.54) + (1500 \times 13.23) + (3000 \times 7.78) = 54{,}725 \text{ contacts}$$

Upper and lower bounds can be calculated by using the estimated standard deviations of the predicted means. The predicted use based on means one standard deviation below the predicted means is

$$(1000 \times 10.94) + (1500 \times 12.81) + (3000 \times 7.20) = 51{,}755 \text{ contacts}$$

The upper bound based on means one standard deviation above the predicted means is

$$(1000 \times 12.14) + (1500 \times 13.65) + (3000 \times 8.36) = 57{,}695 \text{ contacts}$$

These calculations provide a guideline for the number of pediatricians required. But suppose the planner wishes to incorporate additional factors into the process. If there are to be multiple clinics, for example, it may be necessary to have a pediatrician at each clinic instead of a scheduled rotation. The method for prediction assumes that the HMO can control the system to smooth the demand for services. If this is not the case, the number of pediatricians will have to be adjusted to account for the peaks and valleys of demand.

These predicted values are based on one established HMO and its experience. Whether this experience can be transferred to another HMO with a different population is unknown, but the information does provide the planner with a starting point.

SUMMARY

In this chapter we examined a response variable with more than two possible outcomes. Because the outcomes were ordered we were able to form a mean score for each subpopulation and to analyze the associations between the mean score and the set of predictor variables. The analysis suggests that young families (subscribers age 25 to 44) with children less than age 5 and with more than one child use pediatricians more frequently than other subgroups. We next demonstrated the use of these results to predict the number of pediatric contacts for a different population.

The major points addressed in this chapter are:

- *The problem:* Estimation of the number of pediatricians necessary for servicing different populations.
- *The Kaiser data:* Data collected from the HMO prototype.
- *Formation of means:* Assign a score to represent each ordered response category and multiply by the category's probability; then sum across categories.
- *The prediction equation:* The first models did not fit the data; the predicted values could not be used.
- *New factor variables:* The original factors did not convey the needed information; with the new factor variables it was possible to obtain an adequate model.
- *A reduced model:* The unimportant factor gender was deleted from the analysis and as a result more of the sample data could be used in the analysis.
- *Use of the model:* The subscriber's age, number of children, and age of youngest child are used to predict the number of pediatric contacts per year.

Chapters 4, 5, and 6 have demonstrated the use of linear functions of the response variable. As a result of the use of linear functions, interpretation of the parameter estimates and the results has not been hopelessly complicated. In Chapter 7 we consider a log-linear model—that is, a model that involves a logarithmic function of the response variable—and relate this logarithmic function to a linear combination of the factor variables.

EXERCISES

6.1.

Table 6.11 presents data from a study of intergenerational status mobility for British and Danish father-son pairs. If one assumes that the steps on the occupational status scale are approximately equal (1 = highest; 5 = lowest),

Table 6.11 Intergenerational Status Mobility Data for British and Danish Father-Son Pairs

(a) British Data

Father's Occupational Status	SON'S OCCUPATIONAL STATUS					
	1	2	3	4	5	Total
1	50	45	8	18	8	129
2	28	174	84	154	55	495
3	11	78	110	223	96	518
4	14	150	185	714	447	1510
5	3	42	72	320	411	848
Total	106	489	459	1429	1017	3500

(b) Danish Data

Father's Occupational Status	SON'S OCCUPATIONAL STATUS					
	1	2	3	4	5	Total
1	18	17	16	4	2	57
2	24	105	109	59	21	318
3	23	84	289	217	95	708
4	8	49	175	348	198	778
5	6	8	69	201	246	530
Total	79	263	658	829	562	2391

Source: Studies by Glass and Svalastoga as reported in Mosteller (1968).

a "net intergenerational mobility score" may be constructed for these data by using the following weighting scheme. The entries in the body of this table are the weights to be applied to the cell proportions in each subpopulation using a block-diagonal **A** matrix.

		Son's Occupational Status				
		1	2	3	4	5
	1:	0	−1	−2	−3	−4
Father's	2:	1	0	−1	−2	−3
Occupational	3:	2	1	0	−1	−2
Status	4:	3	2	1	0	−1
	5:	4	3	2	1	0

This weighting scheme treats the frequency data as a (2×25) table and produces a 2-DF function vector

$$F = \begin{bmatrix} \bar{X}_B \\ \bar{X}_D \end{bmatrix}$$

Here $F = Ap$ and

$$A = \begin{bmatrix} A^* & 0 \\ 0 & A^* \end{bmatrix}$$

where A^* is of size (1×25). The function will be equal to zero provided that the changes in status mobility in one direction offset the changes in the other (net difference of zero). If the function is not equal to zero, the absolute value of the function is the average intergenerational change in status. The sign of the mobility score indicates the direction of change: A positive sign means that the son's status is on average greater than the father's; a negative sign means that the son's status is on average less than the father's.

a. Analyze the British-Danish data and test the hypotheses that (1) the net intergenerational mobility score is zero and (2) the net intergenerational mobility score between countries is equal. Write a paragraph summarizing your findings.

b. What assumptions must you make about the data to use this function?

c. What alternative functions would you propose for these data?

6.2.

Williams and Grizzle (1970) examined data reported by Bahr (1969) concerning the consumption of alcoholic beverages by people living in different locations and having different times in residence:

Location and Time in Residence (Years)	CONSUMPTION OF ALCOHOLIC BEVERAGES			
	Light and Abstention	*Moderate*	*Heavy*	*Total*
Bowery				
0	25	21	26	72
1–4	21	18	23	62
5+	20	19	21	60
Camp				
0	29	27	38	94
1–4	16	13	24	53
5+	8	11	30	49
Park Slope				
0	44	19	9	72
1–4	18	9	4	31
5+	6	8	3	17

a. Form a mean consumption score.

b. Analyze the relationship between the mean consumption score and the location and time in residence.

c. Form another mean score for alcoholic consumption by using different but still reasonable weights and reanalyze the data. Do you observe any important differences between the two analyses? Discuss how the choice of weights may affect the conclusions one derives.

6.3.

Reanalyze the Lehnen and Reiss data from the National Crime Survey presented in Exercise 5.3, this time using a mean score function. Does the choice of function affect your substantive conclusions? Which function is best for these data: the proportion or mean score? Explain.

APPENDIX: GENCAT INPUT

Frequency Data from Table 6.2

The following listing is for the GENCAT control cards that reproduce the analysis based on Table 6.2.

```
     5     2     1           1          CHAP 6: MEAN SCORES AND HMO DATA, PART I
    16     7                             (7F3.0)
     1
  0.   1.    2.    3.    5.    8.    20.
   2    7    1    2    4    1    9
   3              6    2   11
   2    1    1    3    6    7   10
        2    2    1    2    6   14
   1    2    2    6    8    4   18
   8    5    6    9   24   30131
   9   16    8   11   26    5   28
  16    4    2    6    3    4    6
   2         2    1    6    3    8
   9    2    3    9   13   11   41
   8    6    4    8    7    4   18
   9    1    2    1    2    1    3
  31    3    3    4    3    3    1
   5    5    7    3    8    1    4
  32    4    6    2    6    5    3
  32    2    2    1    1         2
        7         1         7            (40F2.0)          MAIN EFF OF CHILD AGE, AGE + SEX
   1 1 1 1 1 1 1 1 1 1 1 1 1 1 1 1
   1 1 1 1                 -1-1-1-1
           1 1 1 1 1 1 1 1-1-1-1-1
   1 1-1-1 1 1 1 1-1-1-1-1 1 1 1-1
   1-1 1-1 1-1-1-1 1-1-1-1 1-1-1 1
   1 1 1 1 1 1   1 1  -1-1   -1-1
           1-1          1-1-1 1-1-1
        8    1    2                      (40F2.0)          AGE OF PARENT
   1
   1
        8    1    1                      (40F2.0)          SEX OF PARENT
             1
        8    1    1                      (40F2.0)          NUMBER OF CHILDREN
             1
        8    1    2                      (40F2.0)          AGE OF YOUNGEST CHILD
                  1
                       1
```

Frequency Data from Table 6.5

The following listing is for the GENCAT control cards that reproduce the analysis based on Table 6.5.

```
     5     2     1           1          CHAP 6: MEAN SCORES AND HMO DATA, PART II
    16     7                             (7F3.0)
     1
  0.    1.    2.    3.    5.    8.    20.
  27    2    2    1         1    1
   2    5    1    3    5         9
   7    2              6    3   14
  19    2    1    3    1    4    6
   2    1    1    2    7    8    9
   4    2    4    1    3    3   12
  20    3    4    6   10    5   22
  30   25   15   26   52  38164
  32    1    1    2    2    1
  20         4    2   11    3   12
  17    9    8   16   17   15   58
  34    5    3    6    4    3    2
  37    9   13    5   14    9   15
  76    1    1         1         2
  23    2    1    1    1         3
   7    2    3    2    5    3    3
```

```
    7    1    6                  (40F2.0)           AGE SEX AND FAM SIZE
1 1 1 1 1 1 1 1 1 1 1 1 1 1 1 1 1
1 1 1-1-1-1 1 1-1-1-1 1 1-1-1-1
1 1 1 1 1 1            -1-1-1-1-1
                1 1 1 1 1-1-1-1-1-1
1  -1 1  -1   -1 1   -1  -1 1  -1
  1-1     1-1 1-1    1-1 1-1    1-1
    8    1    1                  (40F2.0)           SEX
  1
    8    1    2                  (40F2.0)           AGE
      1
        1
    8    1    2                  (40F2.0)           FAMILY SIZE
           1
             1
```

Frequency Data from Table 6.8

The following listing is for the GENCAT control cards that reproduce the analysis based on Table 6.8.

```
    5    2    1         1        CHAP 6: MEAN SCORES AND HMO DATA, PART III
    9    7                       (7F3.0)
    1
0.   1.   2.   3.   5.   8.   20.
   4   8   2   5  10   8  19
   3   2   2   1   8   8  25
   3   2   4   7  14   7  26
  17   7   9  18  37  41172
  17  22  12  19  33   9  46
  25   5   4   7   5   5   9
  63   5   5   5   4   3   3
   7   5   8   4  10   3   6
  40   6   8   4   9   6   4
    7    1    6                  (40F2.0)           AGE, # CHILDREN, YOUNGEST
1 1 1 1 1 1 1 1 1 1
1 1            -1-1-1-1
      1 1 1 1-1-1-1
1-1 1-1-1-1 1-1-1
1 1 1 1   -1-1   -1
           1-1-1 1-1
    8    1    2                  (40F2.0)           AGE OF PARENT
  1
    1
    8    1    1                  (40F2.0)           NUMBER OF CHILDREN
      1
    8    1    2                  (40F2.0)           AGE OF YOUNGEST CHILD
      1
        1
```

7
Log-Linear Models

The three preceding chapters have all used models in which the response variables were probabilities (Chapters 4 and 5) or a linear combination of probabilities (Chapter 6). In this chapter we consider a model in which the response function involves the natural logarithm of the response variable. The particular form of the logarithmic function that we will use is the *logit*. The logit function is defined as the natural logarithm of π divided by $1 - \pi$:

$$\text{logit}(\pi) = \ln\left(\frac{\pi}{1 - \pi}\right)$$

One rationale for this transformation is presented in this chapter. After performing this transformation, we analyze the logit in the same fashion as the linear functions in Chapters 4, 5, and 6.

Since we are using the log-linear model, it is possible to compare the WLS method with the maximum likelihood approach. We could not make this comparison in earlier chapters because computer programs for the maximum likelihood approach have been developed only for use with log-linear models.

ANALYZING AIR POLLUTION EFFECTS

Much concern has been expressed about the possible effects of air pollution on the nation's health. Severe pollution outbreaks (in Donora, Pennsylvania and London, England, for example) have demonstrated quite clearly that health

98

problems can be caused by large-scale contamination of the air. In many other situations, however, the relationship is not so obvious.

Many studies have been undertaken to investigate the relation between health status and air pollution, but it has been difficult to demonstrate that link conclusively due to the possible confounding effects of other variables. Some of these confounding variables are smoking status, age, sex, ethnicity, and occupation. Additional problems are caused by the lack of a universal measure of health status and by our ignorance of the air pollution process itself. What pollutants should we attempt to measure? Should we be concerned about the possible interaction of conditions? The following study is an attempt to examine the ambient air pollution problem while controlling for a number of important confounding variables. For a detailed presentation of the study, see McClimans (1978).

THE DATA

The San Jacinto Lung Association in Houston, Texas, conducts screening programs designed to detect individual respiratory problems. The screening procedure consists of three parts: (1) a standard chest X-ray, (2) a questionnaire to obtain the medical history and information on coughing or breathing difficulties, and (3) a pulmonary function test (PFT) using a 10-liter spirometer. For purposes of this chapter we will focus on the pulmonary function test. Measurements calculated from the pulmonary function test are FVC (forced vital capacity), $FEV_{1.0}$ (forced expiratory volume in the first second), and the $FEV_{1.0}/FVC$ ratio. The test results are standardized according to height, sex, and age of the respondent and then ranked as normal, borderline, or abnormal according to the number of standard deviations of the result from the predicted value. We will be using this measure as our proxy for health status. If air pollution affects health status in some unexpected way, however, we would probably not detect this relationship by using only the pulmonary function test.

Data from 41 air sampling stations in the Houston–Galveston area were available during the 1974–1975 study period. The information was incomplete for certain pollutants, while for others the measurements were very close to zero. After eliminating pollutants with either incomplete data or small values, four pollutants remained: total suspended particulates, copper, sulfur dioxide, and lead.

Based on the measurements from 41 sampling stations, contour maps were drawn for each of these pollutants. Since the maps showed that the four pollutants differed in their dispersion patterns, we decided not to combine the pollutants into a single index but rather to treat them separately. In this chapter we report only on the results for lead.

The choice of lead is unusual because its effects are reflected primarily by the nervous system, the system for forming blood cells, and the kidneys, but not by the respiratory system. There are two reasons why we suspect that lead, as mapped in this study, may show an association with performance on the pulmonary function test. The first reason is that in the 1974–1975 period automobile emissions served as a major source of lead pollution, and other components of this emission include nitrogen dioxide and complex hydrocarbons, which do have adverse effects on the respiratory system. The second reason involves the process used in mapping the pollutants. Since it was assumed that the pattern of pollutant dispersal is continuous between adjacent contour lines, values between sampling stations were determined by interpolation. Although the mapped dispersion pattern of lead closely resembles that of the lighter hydrocarbons, its *concentration* declines rapidly with distance from the source. For two reasons, then, we suspect that lead may be serving as a proxy for one of the other automobile emission components associated with respiratory problems.

DEFINING THE VARIABLES

The relevant variables include a measure of health status, a variable reflecting pollution levels, and two important confounding variables:

1. Health status (response variable): normal results on the PFT/borderline or abnormal results on the PFT
2. Air pollution (factor variable): high levels of lead in the air at the individual's residence/low levels of lead in the air at the individual's residence
3. Age (confounding variable): less than 40/40 to 59/60 or over
4. Smoking status (confounding variable): *never* smoked/*former* smoker/*light* smoker (has smoked 20 cigarettes or less per day for 10 years or less)/*heavy* smoker (has smoked more than 20 cigarettes per day for less than 10 years or has smoked for more than 10 years)

The few people who smoked pipes or cigars but not cigarettes were deleted from the study.

The individual data used in this chapter were collected by the San Jacinto Lung Association during 1974–1975 from 3006 employees of nine companies in the Houston–Galveston area. For this chapter we consider the data from the largest single company in order to eliminate the possible confounding effect of plant location. We reduce the sample size by eliminating blacks and asthmatics from the study, because the norms used in the PFT standardization process appear to be inappropriate for them. Since there were relatively few people over 40 (245 people in the plant studied)—few in the sense that they will be categorized by the three other variables in the study—we also eliminate people over 40 from the study. Hence age is not a factor in our analysis. We treat the final data set of 479 people (Table 7.1) as a one-response and two-factor problem.

Table 7.1   PFT Results for People Under Age 40 by Smoking Status and Lead Level in Ambient Air

Lead Level and Smoking Status[a]	PFT		Logit (*Normal*)
	Normal	*Not Normal*	
High lead level			
Never	33	3	2.40
Former	12	2	1.79
Light	21	2	2.35
Heavy	16	3	1.67
Low lead level			
Never	160	4	3.69
Former	49	6	2.10
Light	75	6	2.53
Heavy	84	3	3.33

[a] Never smoked; former smoker; light smoker (has smoked 20 cigarettes or less per day for 10 years or less); heavy smoker (has smoked more than 20 cigarettes per day for less than 10 years or has smoked for more than 10 years).

Although we have eliminated the company's location as a confounding variable, we have not considered the exposure in the workplace. No measurements of the air in the workplace were available, but a question about exposure to gases, dust, or fumes on the job had been asked. An examination of the link between this question and performance on the pulmonary function test did not reveal a strong relationship.

A RATIONALE FOR THE LOGIT

A pioneer in the use of the logit function is Berkson (1944, 1946, 1953), who suggested the logit as an alternative to the standard analysis in bioassay. In bioassay the goal is to determine the potency of a stimulus, which is frequently estimated by the *dose* at which 50 percent of the subjects show a response. Often the response of interest is rare, which means that the probability of "success" is less than 0.20. The independent variable in the dose-response analysis is the logarithm of the dose administered as the stimulus. Berkson provided theoretical justification for the use of the logit in bioassay as well as showing that it was easier to calculate than the standard approach. Moreover empirical work has shown that the logit and the standard approach give almost the same results.

Dyke and Patterson (1952) were among the first to use the logit in a non-bioassay application. They considered data with multiple independent variables that were categorical; they did not focus on rare events. This type of application abounds in the literature today. Goodman (1972), for example, applies the logit function to nonrare events in a sociological setting.

Our rationale for using the logit here is that it allows us to test the hypothesis of no multiplicative association between lead exposure and health status, an interesting and important hypothesis, and to examine the interactions among the factors. Because the probability of an abnormal PFT response is less than 0.20, the logit is an especially appropriate choice of function. Above all, it avoids potential mathematical difficulties associated with the use of the simple probability function for rare events, as discussed in Chapter 2.

Table 7.2 presents the table of probabilities for a two-factor, one-response situation in which all three variables have two levels. Note that the logit function requires that there be only two levels for the response variables.

The hypothesis of no multiplicative association among the factors uses ratios of probabilities to determine whether the probabilities of the columns depend on the level of the row variable. If we focus on the first two rows in Table 7.2, the hypothesis of no multiplicative association between the response variable and factor B within level 1 of factor A is given by

$$H_0: \quad \frac{\pi_{11}}{\pi_{12}} = \frac{\pi_{21}}{\pi_{22}} \qquad \text{or} \qquad H_0: \quad \frac{\pi_{11}\pi_{22}}{\pi_{12}\pi_{21}} = 1 \qquad (7.1)$$

The first two rows in Table 7.2 will have the same distribution of the probabilities if the ratio of π_{11} to π_{12} is equal to the ratio of π_{21} to π_{22}.

If there is no multiplicative association among the three variables, the relationship observed between the response variable and factor B should not depend on the levels of A. In other words, the relationship between the response

Table 7.2 Probabilities for a Two-Factor, One-Response Study

FACTOR		RESPONSE		
A	B	*1*	*2*	*Total*
1	1	π_{11}	π_{12}	1
	2	π_{21}	π_{22}	1
2	1	π_{31}	π_{32}	1
	2	π_{41}	π_{42}	1

variable and factor B should be the same within level 1 of A as it is in level 2 of A, which can be expressed as

$$H_0: \frac{\pi_{11}\pi_{22}}{\pi_{12}\pi_{21}} = \frac{\pi_{31}\pi_{42}}{\pi_{32}\pi_{41}} \quad \text{or} \quad H_0: \frac{\pi_{11}\pi_{22}\pi_{32}\pi_{41}}{\pi_{12}\pi_{21}\pi_{31}\pi_{42}} = 1 \tag{7.2}$$

This is the hypothesis of no multiplicative association between the two factor variables in relationship to the dependent variable.

Let us now examine the logistic linear model in order to see its equivalence with Expression (7.2). Recall that the logit, $\ln[\pi/(1 - \pi)]$, is the dependent variable in a model that contains two factors A and B—that is,

$$\ln\left(\frac{\pi_{11}}{\pi_{12}}\right) = \beta_0 + \beta_1 + \beta_2$$

$$\ln\left(\frac{\pi_{21}}{\pi_{22}}\right) = \beta_0 + \beta_1 - \beta_2$$

$$\ln\left(\frac{\pi_{31}}{\pi_{32}}\right) = \beta_0 - \beta_1 + \beta_2 \tag{7.3}$$

$$\ln\left(\frac{\pi_{41}}{\pi_{42}}\right) = \beta_0 - \beta_1 - \beta_2$$

where β_1 represents the differential effect of factor A and β_2 represents the differential effect of factor B. Note that there is no interaction term between factors A and B in this model. The goodness-of-fit test for this model measures whether there is any interaction between factors A and B in relation to the dependent variable, and this is the test of no two-way interaction between the two factor variables and the response variable.

To demonstrate the equivalence of Expressions (7.2) and (7.3), we first assume that the model fits. Taking the exponential of each side of the four equations in (7.3) yields

$$\frac{\pi_{11}}{\pi_{12}} = e^{\beta_0}e^{\beta_1}e^{\beta_2}$$

$$\frac{\pi_{21}}{\pi_{22}} = e^{\beta_0}e^{\beta_1}e^{-\beta_2}$$

$$\frac{\pi_{31}}{\pi_{32}} = e^{\beta_0}e^{-\beta_1}e^{\beta_2} \tag{7.4}$$

$$\frac{\pi_{41}}{\pi_{42}} = e^{\beta_0}e^{-\beta_1}e^{-\beta_2}$$

Substitution of (7.4) into (7.2) yields

$$\frac{\pi_{11}\pi_{22}\pi_{32}\pi_{41}}{\pi_{12}\pi_{21}\pi_{31}\pi_{42}} = \frac{e^{\beta_0}e^{\beta_1}e^{\beta_2}e^{\beta_0}e^{-\beta_1}e^{-\beta_2}}{e^{\beta_0}e^{\beta_1}e^{-\beta_2}e^{\beta_0}e^{-\beta_1}e^{\beta_2}} = 1 \tag{7.5}$$

Because the numerator and denominator contain identical terms, this exercise shows that if the logit model *without* the interaction term fits the data, there is no multiplicative association in the sense of Equation (7.2). If the logit model contains an interaction term β_3, Equation (7.5) yields the following:

$$\frac{\pi_{11}\pi_{22}\pi_{32}\pi_{41}}{\pi_{12}\pi_{21}\pi_{31}\pi_{42}} = e^{4\beta_3}$$

which does not equal 1.00 unless β_3 equals 0.00. We would therefore conclude that multiplicative association exists because Equation (7.2) is no longer true.

We have shown that if there is interaction between the two factors and the dependent variable in the logit model, three-way multiplicative association exists. If there is no interaction between the factors in the logit model, there is no three-way multiplicative association. To complete the demonstration of equivalence, we must show that if Equation (7.2) holds, this implies that the model in Equation (7.3) fits the data.

We begin by assuming that Equation (7.2) holds, or

$$\frac{\dfrac{\pi_{11}}{\pi_{12}} \Big/ \dfrac{\pi_{21}}{\pi_{22}}}{\dfrac{\pi_{31}}{\pi_{32}} \Big/ \dfrac{\pi_{41}}{\pi_{42}}} = 1$$

Taking the natural logarithm of this equation gives

$$\ln\left(\frac{\pi_{11}}{\pi_{12}}\right) - \ln\left(\frac{\pi_{21}}{\pi_{22}}\right) - \ln\left(\frac{\pi_{31}}{\pi_{32}}\right) + \ln\left(\frac{\pi_{41}}{\pi_{42}}\right) = 0 \tag{7.6}$$

For convenience, label $\ln(\pi_{i1}/\pi_{i2}) = a_i$. From Equation (7.6) we see that

$$a_1 - a_2 = a_3 - a_4 \tag{7.7}$$
$$a_1 - a_3 = a_2 - a_4 \tag{7.8}$$

and

$$a_1 + a_4 = a_2 + a_3 \tag{7.9}$$

We will let the constant difference in Equation (7.7) be $2\beta_2$; the constant difference in Equation (7.8) will be called $2\beta_1$; the constant sum in Equation (7.9) can be called $2\beta_0$. Note that we can name these quantities arbitrarily, because the names we assign do not affect the demonstration. We now have four linear equations in four unknowns (a_i):

$$a_1 - a_2 = 2\beta_2 \qquad (7.10)$$

$$a_1 - a_3 = 2\beta_1 \qquad (7.11)$$

$$a_1 + a_4 = 2\beta_0 \qquad (7.12)$$

$$a_1 - a_2 - a_3 + a_4 = 0 \quad \text{(from Equation 7.6)} \qquad (7.13)$$

From these equations we know the following:

$$a_2 = a_1 - 2\beta_2$$

$$a_3 = a_1 - 2\beta_1$$

$$a_4 = -a_1 + 2\beta_0$$

Substituting these into Equation (7.13) yields

$$a_1 - (a_1 - 2\beta_2) - (a_1 - 2\beta_1) + (-a_1 + 2\beta_0) = 0$$

or

$$a_1 = \beta_0 + \beta_1 + \beta_2$$

and

$$a_2 = a_1 - 2\beta_2 = \beta_0 + \beta_1 - \beta_2$$

and

$$a_3 = a_1 - 2\beta_1 = \beta_0 - \beta_1 + \beta_2$$

and finally

$$a_4 = -a_1 + 2\beta_0 = \beta_0 - \beta_1 - \beta_2$$

Hence if Equation (7.2) is true, Equation (7.3) also holds.

Equation (7.3) is the linear model form of the multiplicative relationship shown in Equation (7.2). The relationship between Equations (7.2) and (7.3) demonstrated above provides one rationale for the use of the logit as a dependent variable: The logit function permits tests for no multiplicative association, as expressed in equation (7.2).

In the next section we analyze the data in Table 7.1 by using a linear model approach and WLS estimation with a logit function. Although the analysis is similar to those presented in Chapters 4 through 6, the interpretation of the effects differs somewhat.

THE LOGIT MODEL

The first step in our analysis is to form the logit (normal PFT response) function by taking the logarithm of the probabilities and then multiplying by the $\mathbf{A}$ matrix shown below. The values of the logit (normal) function for each subpopulation are reported in Table 7.1.

$$\mathbf{A}\ln(\mathbf{p}) = \begin{bmatrix} 1 & -1 & 0 & 0 & 0 & 0 & 0 & 0 & 0 & 0 & 0 & 0 & 0 & 0 & 0 \\ 0 & 0 & 1 & -1 & 0 & 0 & 0 & 0 & 0 & 0 & 0 & 0 & 0 & 0 & 0 \\ 0 & 0 & 0 & 0 & 1 & -1 & 0 & 0 & 0 & 0 & 0 & 0 & 0 & 0 & 0 \\ 0 & 0 & 0 & 0 & 0 & 0 & 1 & -1 & 0 & 0 & 0 & 0 & 0 & 0 & 0 \\ 0 & 0 & 0 & 0 & 0 & 0 & 0 & 0 & 1 & -1 & 0 & 0 & 0 & 0 & 0 \\ 0 & 0 & 0 & 0 & 0 & 0 & 0 & 0 & 0 & 1 & -1 & 0 & 0 & 0 & 0 \\ 0 & 0 & 0 & 0 & 0 & 0 & 0 & 0 & 0 & 0 & -1 & -1 & 0 & 0 \\ 0 & 0 & 0 & 0 & 0 & 0 & 0 & 0 & 0 & 0 & 0 & 1 & -1 \end{bmatrix} \begin{bmatrix} \ln(p_{11}) \\ \ln(p_{12}) \\ \vdots \\ \ln(p_{82}) \end{bmatrix}$$

The next step is to examine the model

$$\text{logit (normal PFT response)}$$

$$= \ln\left(\frac{p}{1-p}\right) = \mathbf{A}\ln(\mathbf{p})$$

$$= \text{constant term} + \text{differential lead level effect}$$
$$+ \text{differential smoking level effect}$$

which in matrix terms is

$$\mathbf{A}\ln(\mathbf{p}) = \mathbf{X}\boldsymbol{\beta}$$

where $\mathbf{X}$ and $\boldsymbol{\beta}$ are

$$\mathbf{X}\boldsymbol{\beta} = \begin{bmatrix} 1 & 1 & 1 & 0 & 0 \\ 1 & 1 & 0 & 1 & 0 \\ 1 & 1 & 0 & 0 & 1 \\ 1 & 1 & -1 & -1 & -1 \\ 1 & -1 & 1 & 0 & 0 \\ 1 & -1 & 0 & 1 & 0 \\ 1 & -1 & 0 & 0 & 1 \\ 1 & -1 & -1 & -1 & -1 \end{bmatrix} \begin{bmatrix} \beta_0 \\ \beta_1 \\ \beta_2 \\ \beta_3 \\ \beta_4 \end{bmatrix}$$

The effects in the model are

$$
\begin{bmatrix} \beta_0 \\ \beta_1 \\ \beta_2 \\ \beta_3 \\ \beta_4 \end{bmatrix} = \begin{bmatrix} \text{constant} \\ \text{differential effect of high lead level} \\ \text{differential effect of never smoked} \\ \text{differential effect of former smoker} \\ \text{differential effect of smoking} \\ <10 \text{ years and } <20 \text{ cigarettes/day} \end{bmatrix}
$$

where we are measuring the lead and smoking effects from the mean. If this model provides a reasonable fit, we can determine whether there is a relationship between air pollution (high and low levels of lead) and health status (PFT response) adjusted for smoking status for people less than 40 years of age who work for the same company.

COMPARISON OF WLS AND ML RESULTS

Logit models permit estimation of the parameters β by using weighted least squares (WLS) or maximum likelihood (ML). Maximum likelihood estimation is not generally available for all classes of functions described in this book, and in this sense the WLS method may be applied to a broader class of problems than the ML approach. Since this problem presents an opportunity to compare parameters and other statistics estimated by two different approaches, we do so to illustrate the similarity of results in this type of situation. The ML statistics reported here were provided by the SAS computer package (Barr, Goodnight, and Sall, 1979).

We first analyze the fit of the main effects model of air pollution and smoking status on health status by WLS and ML estimation. The statistics for testing the fit of the model are WLS $X^2_{\text{GOF}} = 2.20$ and ML log-likelihood $X^2_{\text{GOF}} = 2.26$. The 3 degrees of freedom for the goodness-of-fit test statistics come from the difference between the number of observed logits (eight) and the number of parameters (five) for the model. The similarity in the test statistics produced by the two methods is not an unusual finding for problems with reasonable sample sizes. Berkson (1968, 1972) and others have investigated the relationship of WLS and ML estimation in a number of different situations. Their findings lead us to expect agreement between the two procedures, and in the large-sample case, WLS and ML produce equivalent estimates. Although the large-sample situation does not apply to three of the subpopulations ($n < 25$), reasonable agreement still exists between the WLS and ML approaches for this problem.

Comparing the goodness-of-fit test statistic to the chi-square value, $\chi^2_{3, 0.75} = 4.11$, we see that both X^2_{GOF} statistics are less than 4.11, indicating a good fit to the observed data. We next examine the parameter estimates and

Table 7.3 Parameter Estimates and Test Statistics for Lead and Smoking Status Effects

Parameter	WLS Parameter Estimates	ML Parameter Estimates	WLS Test Statistic	ML Test Statistic	DF
Constant	2.42	2.46			
Lead					
High	−0.44	−0.42 ⎫	4.37	4.29	1
Low	0.44	0.42 ⎭			
Smoking status					
Never	0.66	0.64			
Former	−0.62	−0.63			
Light	−0.16	−0.15			
Heavy	0.12	0.14			

investigate whether the air pollution variable (lead) is an important factor relating to health status (PFT response). Since we are only interested in the lead test and not the smoking effects, we perform the test of hypothesis at the $\alpha = 0.05$ level.

Both the WLS and ML parameter estimates and statistics for the test of hypothesis of no lead effect are presented in Table 7.3. Note the close agreement between both sets of parameter estimates and the test statistics. Comparison of either value (4.37 or 4.29) to 3.84 ($\chi^2_{1,0.95}$) causes us to reject the hypothesis of no lead effect. The parameter estimate for the high lead level has a negative value of -0.44. This means that the dependent variable is decreased for the high level of lead category and is increased for the low level of lead category. Recall that the dependent variable is the logarithm of {probability of a normal PFT result divided by probability of a nonnormal result}. If the probability of nonnormal results increases, the ratio of the probabilities of normal results to nonnormal decreases—making the logit smaller. Therefore the high levels of lead are associated with a higher proportion of nonnormal PFT results.

AN ODDS-RATIO INTERPRETATION

We gain another perspective on the interpretation of the parameter estimates if we consider the exponential transformation of the model. The original model is

$$\ln\left(\frac{p_{i1}}{1 - p_{i1}}\right) = \beta_0 + \sum_{j=1}^{4} x_{ij}\beta_j \quad i = 1, 2, \ldots, 8$$

where the x_{ij} are the elements in the **X** matrix. The exponential of this equation is

$$\frac{p_{i1}}{1 - p_{i1}} = e^{\beta_0} e^{x_{i1}\beta_1} e^{x_{i2}\beta_2} e^{x_{i3}\beta_3} e^{x_{i4}\beta_4} \qquad i = 1, 2, \ldots, 8$$

If there is no difference in the ratio of the probabilities among the subpopulations, then β_1 through β_4 are zero and e^{β_0} represents the weighted average value of p_{i1}/p_{i2}. The ratio of p_{i1}/p_{i2} can be called the *odds of a normal PFT* for the *i*th population. If the subpopulations make a difference, the term $e^{x_{ij}\beta_j}$ shows the impact of the *j*th parameter on the odds function. In this problem,

$$b_0 = 2.42 \qquad b_1 = -0.44 \qquad b_2 = 0.66$$
$$b_3 = -0.62 \qquad b_4 = -0.16$$

The predicted odds of a normal PFT for a subject who is a former smoker exposed to high levels of lead pollution is

$$\text{Predicted odds} = e^{b_0} e^{b_1} e^{b_3}$$
$$= e^{2.42} e^{-0.44} e^{-0.62}$$
$$= 11.246 \times 0.644 \times 0.538$$
$$= 3.90$$

We see from this expression that, overall, the odds of a normal PFT are about 11 to 1 ($e^{2.42} = 11.246$). If a former smoker is exposed to high levels of lead, however, this reduces the odds of a normal PFT to about 4 to 1 (3.90). If lead had not had a significant effect, then $b_1 = 0$ and $e^0 = 1$, which means that it would have no effect on the predicted odds of a normal PFT. This, of course, is not the result; in fact, high lead exposure reduces the predicted odds of a normal PFT to about 60 percent (0.644) of normal.

LEAD—A SIGNIFICANT FACTOR?

This study has examined the relationship between ambient air pollution (focusing on lead) and pulmonary function test results while controlling for smoking status. The results reported here are for employees of a single corporation in Houston. Control for in-plant exposure to pollutants is not reported here; however, other analyses indicate no relationship between in-plant exposure to pollutants and performance on the pulmonary function test.

The logit model that we fitted to the data included the main effects of lead and smoking status. The test for the goodness of fit did not provide evidence to cause us to reject this model. Based on the acceptance of this model we then tested for no lead effect but rejected the hypothesis of no lead effect on the logit scale. The high level of lead category is associated with a higher proportion of people having a nonnormal PFT result than for those in the low level of lead category.

The significant lead effect is a surprising result in view of the effects of lead reported in other studies. Lead air pollution has not been shown to have a negative impact on the respiratory system. A possible explanation for the discrepancy in findings springs from a consideration of some of the sources of lead air pollution. During the 1974–1975 period, automobile emissions were a major source of lead air pollution. It is possible that we are observing the effects of other pollutants in the automobile exhaust and not the effects of lead. Nitrogen dioxide is one component of automobile exhaust that has a proven adverse respiratory effect. Examination of the maps for lead air pollution lend support to this argument, since the areas of highest lead concentration are also heavy traffic areas. This pattern suggests that there is a need for further research on the impact of automobile emissions on the nation's health.

SUMMARY

This chapter introduced the log-linear model. We demonstrated the use of the logit in a problem that examined the relationship between air pollution and health with controls for important confounding variables. The use of the log-linear model allowed us to compare the results from using the weighted least squares procedure with the maximum likelihood approach. The results were very close: The test statistics for examining the fit of the model were 2.20 for WLS and 2.26 for ML. This close agreement was not surprising, since the procedures are equivalent in the large-sample case and numerous other comparisons have also provided evidence for close agreement in the small-sample case.

Since the main effects model fits, we were able to test the hypothesis of no lead effect on the pulmonary function test. We rejected the hypothesis of no lead effect and again there was close agreement between the WLS and ML test statistics. We found that people in areas of high lead concentration were associated with poor performance on the pulmonary function test when the data are examined on the logit scale. Since lead itself has no known adverse effect on the respiratory system, this result may mean that we are observing the impact of another pollutant found with lead.

The key points discussed in this chapter are:

- *The problem:* Investigate the effects of lead air pollution on PFT while controlling for smoking status.

- *The data:* From the San Jacinto Lung Association and 41 air sampling stations.
- *Definition of variables:* Response variable—PFT results; factor variables—lead pollution and smoking status.
- *Rationale for the logit:* The study of rare events to estimate odds ratio.
- *The logit model:* The main effects model fits; the odds-ratio may be estimated by the product of the exponential of the mean and main effects.
- *WLS and ML results:* Very similar estimates because of moderate sample size.
- *Results:* Lead is significantly related to PFT results; perhaps lead is a proxy for another pollutant.

This chapter concludes Part II, which presented the analysis of simple functions based on the cell frequencies of a contingency table. Part II introduced three different functions—the proportion, the mean score (an example of weighted cell frequencies), and the logit (an example of log-linear analysis). In Part III we expand on the use of these simple functions and add complexity to the analysis in two different ways. Chapters 8, 9, and 10 illustrate the analysis of more than one function formed in each subpopulation. This is called *multi-response analysis.* Chapters 8 and 10 demonstrate multiresponse analyses of proportions and mean scores.

Complexity is also introduced by the choice of functions. Chapters 9 and 11 analyze *complex functions* of the frequencies. In Chapter 9 we introduce the analysis of the rank correlation coefficient gamma, and in Chapter 11 we present an adjusted probability of survival based on follow-up life table data.

EXERCISES

7.1.

One example of the exercise of discretion in the administration of criminal justice is the prosecutor's decision either to try a criminal case or to dismiss the charges (nolle prosequi). Table 7.4, taken from a study by Lehnen and Koch (1974c), summarizes the prosecutor's decision in minor criminal cases by offense charged at arrest, the county where the arrest occurred, and the race of the defendant. Determine whether there is evidence that the location of an arrest and the defendant's race are related to a prosecutor's decision to try the case by performing the following tasks.

a. What function is best suited to characterize the nolle prosequi rate? Would "proportion nolle prosequi" be a suitable function? Explain.

b. Construct the logit(nolle prosequi) function for each of these 20 subpopulations by hand. Examine these values and guess whether the offense at

Table 7.4   Prosecutor's Decision to Prosecute Misdemeanors in North Carolina by Offense Charged at Arrest, County of Arrest, and Defendant's Race

Offense[a]	County[b]	Race[c]	PROSECUTOR'S DECISION	
			Trial	*Nolle Prosequi*
A	D	B	41	4
A	D	W	55	2
A	O	B	15	1
A	O	W	16	0
V	D	B	20	3
V	D	W	15	1
V	O	B	10	5
V	O	W	6	7
P	D	B	17	2
P	D	W	15	2
P	O	B	16	3
P	O	W	9	0
T	D	B	6	1
T	D	W	19	2
T	O	B	18	1
T	O	W	26	1
S	D	B	35	0
S	D	W	92	3
S	O	B	23	2
S	O	W	114	7

[a] Offense: A = alcohol-related; V = personal violence; P = property; T = traffic (except speeding); S = speeding.
[b] County: D = Durham; O = Orange.
[c] Race: B = black; W = White.

arrest, the county of arrest, and the defendant's race are related to the decision not to prosecute.

c. Fit a main effects model to these logit functions containing a general mean term, four offense effects, one county effect, and one race effect using the GENCAT program. Evaluate the fit of the model by examining the goodness-of-fit X^2 test and the residuals. Does the main effects model provide an adequate fit? Why or why not?

d. It is generally recognized that the charge at arrest is one of the primary reasons for variation in the processing of cases. Fit a modular model to these data that makes the effects of county and race conditional on offense. Complete the following table:

Description	b_i	DF	X^2
Mean for alcohol (A)			
County effect: A			
Race effect: A			
Mean for violence (V)			
County effect: V			
Race effect: V			
Mean for property (P)			
County effect: P			
Race effect: P			
Mean for traffic (T)			
County effect: T			
Race effect: T			
Mean for speeding (S)			
County effect: S			
Race effect: S			

e. Construct the following tests:

1. A 5-DF test that all race effects equal zero
2. A 5-DF test that all county effects equal zero
3. Ten 1-DF tests for the conditional county and race effects
4. Four 1-DF tests between offense mean terms

Do these tests suggest terms that may be omitted from a reduced model? Interpret each test. Are there other tests you would suggest?

f. Fit a reduced model based on the tests done in part (e) above. Evaluate the fit of your reduced model. Is it a better model than the main effects model fitted in part (c) above? Explain what you mean by "better."

g. What substantive conclusions might you make from these data about the prosecutor's decision to try misdemeanors? How far can you generalize your results? What caveats would you place on your interpretation?

7.2.

Goodman (1972) analyzes data reported originally by Stouffer and others (1949) on the preferences of draftees for training camp location based on race,

Table 7.5   Preference of Soldiers for Camp Location by Race, Region of Origin, and Location of Present Camp

Race	Region of Origin	Location of Present Camp	Number of Soldiers in North	Preferring Camp in South
Black	North	North	387	36
Black	North	South	876	250
Black	South	North	383	270
Black	South	South	381	1712
White	North	North	955	162
White	North	South	874	510
White	South	North	104	176
White	South	South	91	869

Source: Goodman (1972).

region of origin, and location of present camp using logit(Northern) function and maximum likelihood estimation (Table 7.5).

a. Analyze these data as a three-factor, one-response problem by using the logit(Northern) function. Fit a saturated model using the following model:

$$
\begin{bmatrix} f_1 \\ f_2 \\ f_3 \\ f_4 \\ f_5 \\ f_6 \\ f_7 \\ f_8 \end{bmatrix} =
\begin{bmatrix}
1 & 1 & 1 & 1 & 1 & 1 & 1 & 1 \\
1 & 1 & 1 & -1 & 1 & -1 & -1 & -1 \\
1 & 1 & -1 & 1 & -1 & 1 & -1 & -1 \\
1 & 1 & -1 & -1 & -1 & -1 & 1 & 1 \\
1 & -1 & 1 & 1 & -1 & -1 & 1 & -1 \\
1 & -1 & 1 & -1 & -1 & 1 & -1 & 1 \\
1 & -1 & -1 & 1 & 1 & -1 & -1 & 1 \\
1 & -1 & -1 & -1 & 1 & 1 & 1 & -1
\end{bmatrix}
\begin{bmatrix} \beta_0 \\ \beta_1 \\ \beta_2 \\ \beta_3 \\ \beta_4 \\ \beta_5 \\ \beta_6 \\ \beta_7 \end{bmatrix}
$$

1. What interpretation may be given to each effect in the model?
2. Test whether each effect equals zero and interpret your results.
3. Compare your parameter estimates (b_i) with the ML results reported by Goodman (1972, table 4, p. 36). Explain why they are similar or different.
4. Compute $b_i^* = e^{b_i}$ for each of the eight parameter estimates ($i = 0, 7$). Compare and contrast the b_i^* to the estimates reported by Goodman (1972, table 4).

b. Fit a reduced model based on the results of hypothesis tests conducted in part (a). Compare your parameter estimates with those reported by Goodman

(1972, table 3). Do your results derived from the WLS method agree with the ML estimates? Explain why or why not.

c. Compute the odds ratio ($e^{\hat{f}_i}$) and the expected frequencies for your reduced model and compare them to the estimates reported by Goodman (1972, table 2).

d. Write a short summary interpreting the b_i, b_i^*, $\hat{f}_i$, odds ratio, and related statistics derived from your reduced model. What factors play an important role in determining preferences for the location of a training camp?

e. The Stouffer data are suitable for analysis using the function "proportion preferring a Northern camp." Reanalyze these data using the proportion function, and fit a reduced model. Interpret your substantive findings and compare them to the logit(Northern) analysis. Which analysis do you prefer for these data? Specify the criteria you used in making your choice.

7.3.

Consider the data presented in Exercise 5.2 (Chapter 5).

a. Why would the use of the logit function be desirable for these data?

b. List the possible factors and the interaction terms for the saturated model and the degrees of freedom associated with each factor and interaction.

c. What is the **X** matrix associated with the terms listed in part (b)?

d. Use GENCAT to fit the saturated model and provide the value of X^2 for all the terms in the saturated model.

e. Examine the results of part (d) and fit a reduced model based on them.

f. Interpret your results. What suggestions would you make to the company and the employees' union?

g. Compare your analysis and conclusions with those you made in Exercise 5.2.

7.4.

One log-linear function used in many applications of categorical analysis is the *log cross-product ratio*. Consider the following 2 × 2 contingency table with cell frequencies a, b, c, and d:

		Characteristic Y: Attitude Toward Republican Candidate[a]	
		+	−
Characteristic X:	Republican	a	b
Vote for President	Democrat	c	d

[a] (+) = favorable view of Republican candidate; (−) = unfavorable view of Republican candidate.

The log cross-product ratio $\ln(\phi_{XY})$ is

$$\ln(\phi_{XY}) = \ln\left(\frac{ad}{bc}\right)$$

or

$$\ln(\phi_{XY}) = \ln(a) - \ln(b) - \ln(c) + \ln(d)$$

The function $\ln(\phi_{XY})$ will be equal to zero when there is no association between characteristics X and Y. It may vary between negative and positive infinity, and the absolute value of the function is a measure of strength of association. A positive value means that a positive association exists between characteristics X and Y; a negative value means that an inverse relationship exists.

Use Table 7.6—1948 presidential election study panel data collected by Lazarsfeld (1948) as reported by Goodman (1973)—to analyze the strength of association between attitudes toward the Republican candidate and voting intention at two points in time.

Table 7.6   Responses of 266 People at Two Points in Time to Questions Pertaining to Voting Intention and Perception of Republican Presidential Candidate

		SECOND INTERVIEW			
Vote:[a]		R	R	D	D
Candidate:[b]		+	−	+	−
FIRST INTERVIEW					
Vote	*Candidate*				
R	+	129	3	1	2
R	−	11	23	0	1
D	+	1	0	12	11
D	−	1	1	2	68

Source: Goodman (1973).
[a] R = Republican; D = Democrat.
[b] (+) = favorable view of Republican candidate;
(−) = unfavorable view of Republican candidate.

a. Form the log cross-product ratio functions for voting intention and attitude toward the Republican candidate for each time period. The (16×1) probability vector is

$$\mathbf{p} = \begin{bmatrix} p_1 \\ p_2 \\ \vdots \\ p_{16} \end{bmatrix}$$

where $p_1 = 129/266$, $p_2 = 3/266, \ldots, p_5 = 11/266$, $p_6 = 23/266, \ldots, p_9 = 1/266$, $p_{10} = 0/266, \ldots, p_{13} = 1/266, \ldots$, and $p_{16} = 68/266$. Note that this formulation of the contingency table treats the data as a no-factor, 4-response problem. This is a multi-response analysis using log-linear functions. Chapter 8 provides a more complete treatment of multi-response analysis.

Form the (2×1) function vector $\mathbf{F}$ using the following transformation:

$$\mathbf{F} = \mathbf{A}_2 \ln[\mathbf{A}_1 \mathbf{p}]$$

where

$$\mathbf{A}_1 = \begin{bmatrix} 1111 & 0000 & 0000 & 0000 \\ 0000 & 1111 & 0000 & 0000 \\ 0000 & 0000 & 1111 & 0000 \\ 0000 & 0000 & 0000 & 1111 \\ 1000 & 1000 & 1000 & 1000 \\ 0100 & 0100 & 0100 & 0100 \\ 0010 & 0010 & 0010 & 0010 \\ 0001 & 0001 & 0001 & 0001 \end{bmatrix} \quad \begin{array}{l} \textit{row label} \\ a:\ \mathrm{R+},\ \text{time 1} \\ b:\ \mathrm{R-},\ \text{time 1} \\ c:\ \mathrm{D+},\ \text{time 1} \\ d:\ \mathrm{D-},\ \text{time 1} \\ a:\ \mathrm{R+},\ \text{time 2} \\ b:\ \mathrm{R-},\ \text{time 2} \\ c:\ \mathrm{D+},\ \text{time 2} \\ d:\ \mathrm{D-},\ \text{time 2} \end{array}$$

and

$$\mathbf{A}_2 = \begin{bmatrix} 1 & -1 & -1 & 1 & 0 & 0 & 0 & 0 \\ 0 & 0 & 0 & 0 & 1 & -1 & -1 & 1 \end{bmatrix} \quad \begin{array}{l} \textit{row label} \\ \ln(\phi_{vc1}) \\ \ln(\phi_{vc2}) \end{array}$$

The resulting function vector is

$$\mathbf{F} = \begin{bmatrix} \ln(\phi_{vc1}) \\ \ln(\phi_{vc2}) \end{bmatrix}$$

b. Test the hypothesis that the strength of association between attitude toward the Republican candidate and voting intention is the same at each time period.

c. Write a paragraph interpreting your results.

d. What other log cross-product ratio functions might be formed from Table 7.6? How would you interpret them in an analysis?

APPENDIX: GENCAT INPUT

The following listing is for the GENCAT control cards that reproduce the analysis described in this chapter.

```
        5   1   1              CHAP 7: ANALYSIS OF LOG-LINEAR FUNCTION
        8   2                  (2F3.0)
   33   3
   12   2
   21   2
   16   3
  160   4
   49   6
   75   6
   84   3
        2
        1   2   8   1          (2F2.0)
01-1
        7   1   5              (8F2.0)          MAIN EFFECTS MODEL
0101010101010101
01010101-1-1-1-1
010000-1010000-1
000100-1000100-1
000001-1000001-1
        8   1   1              (5F1.0)          LEAD EFFECT
01
```

PART III

Advanced Applications of the WLS Approach

8

Multiple Response Functions

Part II presented problems involving a *single-response* function within each subpopulation. The goal was to summarize the variation in this function *across* subpopulations. In Part III we consider both complex single-response functions and also *multiple* functions within each subpopulation. In the latter or *multiresponse* situation, one objective is to test hypotheses about the functions *within* subpopulations.

Multiresponse problems arise from many design situations. In *survey panel* designs, the same subjects are measured at least two different times—for example, at different points during an election campaign (Lehnen and Koch, 1974b). Multiresponse situations also arise whenever multiple measurements are taken on the same subject or when patients respond to multiple stimuli. Grizzle, Starmer, and Koch (1969) analyze data arising from a situation in which each patient is treated with three drugs; Koch and Reinfurt (1971) examine the multiple responses of people who have been exposed to a number of stimuli in a psychological experiment.

Although there is a rich array of approaches to the analysis of multiresponse data, we will illustrate one approach that focuses on the information contained in the row and column margins and is especially useful for the problems common to public health and social policy. Chapters 9 and 10 present alternative techniques for the analysis of multiresponse situations.

EXAMINING TRUST IN GOVERNMENT

The growth of government-sponsored social welfare programs and the development of social survey methods has spawned a host of studies to measure attitudes toward public institutions, policies, and programs. During the social upheavals of the late 1960s and early 1970s, government institutions took special interest in the attitudes of citizens toward them and their performance. In 1973, for example, a committee of the United States Senate commissioned a special survey and released its results under the title *Confidence and Concern: Citizens View American Government* (U.S. Senate, 1973). University-based survey operations, such as the University of Michigan Survey Research Center and the University of Chicago National Opinion Research Center, also have conducted periodic studies to reveal public attitudes toward government.

The problem discussed here illustrates the type of data produced by social surveys. It is taken from the Southeast Regional Survey (SERS) Project conducted by the University of North Carolina. (See Lehnen, 1976, for details.) During the academic year 1968–1969, a national sample of adult Americans was interviewed to determine their attitudes toward government institutions, officials, and policies. Included in the questionnaire were three agree/disagree questions pertaining to trust in government institutions:

(The President/the Supreme Court/United States Senators)
can be trusted to do what is good for the people.

Subjects were given the opportunity to express an "agree," "disagree," "not sure" (ambivalent), or "have no opinion" response. Table 8.1 presents the distribution of responses to the three trust questions classified by respondent's race and income in a two factor, three response format.

The problem central to this chapter is to determine the level of trust expressed toward each of the three institutions (President, Supreme Court, Senate) and how these levels of trust are affected by the respondent's race (white, black) and income (low, medium, high). The hypotheses of interest may be grouped as follows:

1. Hypotheses concerning the interrelationship of the three trust questions
2. Hypotheses concerning the differences in trust across race by income subpopulations
3. Hypotheses concerning the interaction of trust question responses with race and income

The first group of hypotheses represents the multiresponse aspects of the problem. Since each subject is classified in Table 8.1 according to the combination of agree/disagree responses, the functions based on these cell frequencies

Table 8.1 Distribution of Responses to Three Trust Questions by Race and Income

Question[a] President:	Court:	Senate:	D D D	D D A	D D N	D A D	D A A	D A N	D N D	D N A	D N N	A D D	A D A	A D N	A A D	A A A	A A N	A N D	A N A	A N N	N D D	N D A	N D N	N A D	N A A	N A N	N N D	N N A	N N N
Race	**Income**																											
White	Low		15	3	1	5	8	4	4	2	3	13	17	5	12	60	10	4	12	8	9	3	2	2	10	2	2	3	7
White	Medium		41	17	4	12	17	0	2	2	1	37	31	10	29	127	15	8	17	13	12	6	5	5	10	5	2	6	12
White	High		35	10	4	21	12	0	2	1	3	30	46	2	18	136	12	8	24	10	11	2	1	4	6	5	3	1	3
Black	Low		12	4	2	0	6	4	4	0	3	8	1	3	3	81	7	6	6	2	2	0	7	10	6	3	3	3	8
Black	Medium		8	1	1	4	9	1	0	2	1	3	2	0	10	30	8	7	1	3	3	2	2	6	5	2	1	1	10

Source: Data adapted from Lehnen and Koch (1974a).

[a] Each respondent was administered three agree/disagree questions of the form "(The President/the Supreme Court/United States Senators) can be trusted to do what is good for the people." D = disagree; A = agree; N = not sure or no opinion.

are dependent, and hypotheses about differences in trust levels *within a sub-population* must take account of these dependencies.

If one ignores the interrelationships of the trust responses and analyzes each trust question separately, then the focus is on the second group of hypotheses. These hypotheses examine the differences in trust *across* subpopulations and thus are similar to the single-response problems presented in previous chapters.

The third group of hypotheses relates to the effects of subpopulation characteristics on the interrelation of the trust responses. Here we are interested in whether different combinations of race and income affect the relationship among the trust responses.

The data presented in Table 8.1 are from a study first reported by Lehnen and Koch (1974a). Unlike data reported in other chapters, which may be viewed as resulting from a simple random sample, these data arose from a complex sample survey design. (Details of the sample design were not available.) For the purpose of illustrating the following analyses we rounded the reweighted frequencies reported by Lehnen and Koch to the nearest whole numbers and will view the rounded data as approximating the results of a simple random sample. The issues involved in reweighting the original trust data are discussed by Lehnen and Koch, and Chapter 12 reviews literature devoted to the problem of estimation from complex sample surveys. We have also excluded the black high-income subpopulation from the analysis because there was insufficient sample size to meet the guidelines (specified in Chapter 2) for the multiple proportion analysis reported below. Approximately 30 cases out of a total sample size of 1500 were excluded.

STRUCTURE OF A MULTIRESPONSE TABLE

In the multifactor, single-response problems examined in Chapters 4 to 7, we analyzed single functions—say, a proportion, a mean score, or a logit—constructed for each subpopulation. Panel (*a*) of Table 8.2 illustrates the probability structure of this type of contingency table. It demonstrates that the underlying cell parameters π_{ij} are constrained such that they sum to 1.00 for each subpopulation. The probabilities of the cells are *conditional* on each subpopulation. When we construct a function across the estimates of the probabilities in each *row* of the table, the row functions will be independent of each other. This fact implies that the structure of the variance-covariance matrix $\mathbf{V}_F$ will be diagonal with estimates of the variance of each function f_i appearing on the diagonal. The off-diagonal estimates of covariance between each pair of functions will be zero.

A multiresponse situation is different from the single-response structure because we are now interested in analyzing more than one function *within* each subpopulation—in other words, we are interested in the *joint* distribution

Table 8.2 Examples of Table Structures for Multifactor, Single-Response, and Multiresponse Designs

(a) *Multifactor, Single-Response Table*

		Response				
		1	2	...	r	Total
	1	π_{11}	π_{12}	...	π_{1r}	1.0
Subpopulations	2	π_{21}	π_{22}	...	π_{2r}	1.0
	$\vdots$	$\vdots$	$\vdots$	...	$\vdots$	$\vdots$
	s	π_{s1}	π_{s2}	...	π_{sr}	1.0

(b) *Single-Subpopulation, Two-Response Table*

Response X:	1		2		...	x		
Response Y:	1 2 ... y		1 2 ... y		...	1 2 ... y		Total

$X \times Y$ response combinations:

$$\pi_{11}\pi_{12}\cdots\pi_{1y} \quad \pi_{21}\pi_{22}\cdots\pi_{2y} \quad \cdots \quad \pi_{x1}\pi_{x2}\cdots\pi_{xy} \qquad 1.0$$

(c) *Panel (b) Reformatted*

		Response Y				
		1	2	...	b	Total
	1	π_{11}	π_{12}	...	π_{1y}	$\pi_{1\cdot}$
Response	2	π_{21}	π_{22}	...	π_{2y}	$\pi_{2\cdot}$
X	$\vdots$	$\vdots$	$\vdots$	$\vdots$	$\vdots$	$\vdots$
	x	π_{x1}	π_{x2}	...	π_{xy}	$\pi_{x\cdot}$
	Total	$\pi_{\cdot1}$	$\pi_{\cdot2}$	...	$\pi_{\cdot y}$	1.0

of the responses. Panels (b) and (c) of Table 8.2 illustrate the case of one subpopulation and two response variables. The response variable X has x levels while response variable Y has y levels. In panels (b) and (c) we have deleted the subpopulation subscript since we are considering one subpopulation. The first subscript of π_{ij} in panels (b) and (c) now refers to the level of response variable X; the second subscript refers to the level of response variable Y.

Panel (c) differs from panel (a) because the sum of the row probabilities is not 1.00. The marginal probabilities $\pi_{i\cdot}$ and $\pi_{\cdot j}$ are simply the sum of the corresponding row or column probabilities. If there are multiple subpopulations, the data may be presented by using the format shown in panel (b) with each row in the table having $x \times y$ entries. The data could also be presented in the format shown in panel (c) except that there would be a table for each subpopulation. The data in Table 8.1 follow the format of panel (b), where each column represents one of the $3 \times 3 \times 3 = 27$ combinations, and there are five subpopulations (rows).

With categorical data, it is possible to construct multiple functions within the same subpopulation when one has only a single response variable as well as in situations where there is more than one. Whenever the analyst constructs *more than one function* within a subpopulation, we will consider it to be a *multiresponse* situation. In the following section we analyze a measure based on the marginal totals: a mean trust score similar to the function discussed in Chapter 6.

MEAN TRUST SCORES

The three response categories disagree (D), agree (A), and not sure/no opinion (N) present several alternatives regarding the choice of function. Some analysts would treat the not sure/no opinion (N) category as a middle category on a psychological continuum bounded by disagree and agree. Using this approach as a guide and assuming an equal distance between categories, we may then assign weights of -1 to disagree, 0 to not sure/no opinion, and $+1$ to agree responses to form a mean trust score. The trust score ranges from -1.00 to 1.00, and a value of zero means that the proportions disagreeing and agreeing are equal. A negative score indicates more disagreement than agreement; a positive score, more agreement than disagreement. This approach is equivalent to taking the difference in the probabilities of agreement and disagreement. This function does not summarize all the ways the marginal trust distributions may vary, but it does capture one important characteristic: the central tendency. It is also possible to examine both the probabilities of agreement and disagreement or other functions, but we shall focus on the mean trust score first.

Three functions are required for each of the five subpopulations. The first function forms the mean trust score for the President, the second function forms the mean for the Supreme Court, and the third function forms the Senate mean. These means may be constructed by using the following **A** matrix:

$$\underset{(15 \times 1)}{\mathbf{F}} = \underset{(15 \times 135)}{\mathbf{A}} \underset{(135 \times 1)}{\mathbf{p}}$$

where **A** is a block-diagonal matrix with $\mathbf{A}^*$, a (3×27) matrix that forms the appropriate linear combinations of the probabilities for each subpopulation, being the entry on the diagonal:

$$
A = \begin{bmatrix} A^* & & & & \\ & A^* & & & \\ & & A^* & & \\ & & & A^* & \\ & & & & A^* \end{bmatrix}
$$

and

$$
\underset{(3 \times 27)}{A^*} = \begin{bmatrix} -1 & -1 & -1 & -1 & -1 & -1 & -1 & -1 & -1 & 1 & 1 & 1 & 1 & 1 & 1 & 1 & 1 & 1 & 0 & 0 & 0 & 0 & 0 & 0 & 0 & 0 & 0 \\ -1 & -1 & -1 & 1 & 1 & 1 & 0 & 0 & 0 & -1 & -1 & -1 & 1 & 1 & 1 & 0 & 0 & 0 & -1 & -1 & -1 & 1 & 1 & 1 & 0 & 0 & 0 \\ -1 & 1 & 0 & -1 & 1 & 0 & -1 & 1 & 0 & -1 & 1 & 0 & -1 & 1 & 0 & -1 & 1 & 0 & -1 & 1 & 0 & -1 & 1 & 0 & -1 & 1 & 0 \end{bmatrix}
$$

Note that A has 15 rows because three means are formed for each of the five subpopulations; it has 135 columns because the probability vector has 135 elements (27 probabilities in each of the five subpopulations). The 15 observed mean trust scores are reported in Table 8.3. We find that all mean scores are

Table 8.3   Observed and Predicted Mean Trust Scores

Race and Income	Question	Observed Mean Trust Score	Predicted Mean Trust Score[a]
White			
Low	President	0.42	0.44
	Court	0.20	0.17
	Senate	0.23	0.25
Medium	President	0.43	0.44
	Court	0.12	0.17
	Senate	0.20	0.25
High	President	0.48	0.44
	Court	0.18	0.17
	Senate	0.26	0.25
Black			
Low	President	0.44	0.44
	Court	0.47	0.43
	Senate	0.38	0.25
Medium	President	0.32	0.44
	Court	0.43	0.43
	Senate	0.22	0.25

[a] Based on model $X_2 \beta_2$.

positive, which indicates a generally trusting response, and that they range from a low of 0.12 for the Supreme Court among medium-income whites to a high of 0.47 for the Supreme Court among low-income blacks.

We wish to examine these means to determine whether relationships exist with the race or income variables—and, if relationships do exist, to be able to express them succinctly. We begin this analysis by considering the following X_1 matrix and β_1 vector:

$$
F = \begin{bmatrix}
1000 & 0000 & 0000 \\
0000 & 1000 & 0000 \\
0000 & 0000 & 1000 \\
1010 & 0000 & 0000 \\
0000 & 1010 & 0000 \\
0000 & 0000 & 1010 \\
1001 & 0000 & 0000 \\
0000 & 1001 & 0000 \\
0000 & 0000 & 1001 \\
1100 & 0000 & 0000 \\
0000 & 1100 & 0000 \\
0000 & 0000 & 1100 \\
1110 & 0000 & 0000 \\
0000 & 1110 & 0000 \\
0000 & 0000 & 1110
\end{bmatrix}
\begin{bmatrix}
\beta_1 \\
\beta_2 \\
\beta_3 \\
\beta_4 \\
\beta_5 \\
\beta_6 \\
\beta_7 \\
\beta_8 \\
\beta_9 \\
\beta_{10} \\
\beta_{11} \\
\beta_{12}
\end{bmatrix}
$$

Model $X_1\beta_1$ follows a simple pattern in which parameters β_1, β_5, and β_9 are constant trust scores for the President, Supreme Court, and Senate respectively for low-income whites. Parameters β_2, β_6, and β_{10} represent the black effect on response to each question, while parameters β_3, β_7, and β_{11} are effects of medium income on each question. Finally, parameters β_4, β_8, and β_{12} are the effects of being white and having high income for each question.

This parameterization uses low-income whites as the reference group for each trust question. The estimated effects for being black, for medium income, and for high income and white may be interpreted as the differences from low-income whites that these conditions create in trusting responses. Model $X_1\beta_1$ uses only 12 of the available 15 degrees of freedom. The three unused degrees of freedom represent the race × income × question interaction that can be estimated. These 3 degrees of freedom will be reflected by the goodness-of-fit test. We are unable to estimate the complete race × income × question interaction because the shortage of sample information on high-income blacks caused the exclusion of this subpopulation from the analysis and the loss of 3 degrees of freedom.

The goodness-of-fit test statistic for $\mathbf{X}_1\boldsymbol{\beta}_1$ provides a test of the hypothesis that no race × income × question interaction exists among the available categories. The test statistic is $X^2_{GOF} = 3.18$ with 3 degrees of freedom, and this is less than $\chi^2_{3,\,0.75} = 4.11$. We thus have an acceptable fit, and we may proceed to test selected hypotheses about the parameters in the model.

Table 8.4 presents the estimated parameters and chi-square test statistics for the black, medium-income, and white high-income effects for each question. With the exception of the black effect for the Supreme Court question, the test statistics are less than the critical value $\chi^2_{1,\,0.99} = 6.63$. These tests show no statistically significant sources of variation in responses to the trust questions resulting from having medium or high income. Race does appear to make a difference, however, depending on the question.

Table 8.4   Estimated Parameters and Selected Tests of Significance Under Model $\mathbf{X}_1\boldsymbol{\beta}_1$

Parameter	Description	b_i	X^2
1	Trust for low-income whites: President	0.46	—
2	Black effect: President	−0.04	0.62
3	Medium-income effect: President	−0.04	0.62
4	White and high-income effect: President	0.02	0.15
5	Trust for low-income whites: Court	0.18	—
6	Black effect: Court	0.28	23.50
7	Medium-income effect: Court	−0.05	0.85
8	White and high-income effect: Court	−0.00	0.003
9	Trust for low-income whites: Senate	0.26	—
10	Black effect: Senate	0.09	2.15
11	Medium-income effect: Senate	−0.08	1.86
12	White and high-income effect: Senate	−0.00	0.002

Since many of the estimates represented by model $\mathbf{X}_1\boldsymbol{\beta}_1$ are not statistically significant at the $\alpha = 0.01$ level, we fit a reduced model $\mathbf{X}_2\boldsymbol{\beta}_2$ that excludes the nonsignificant income and race effects for President and Senate. The new model $\mathbf{X}_2\boldsymbol{\beta}_2$ includes the three constant terms and a race effect for the Supreme Court question:

$$
F = \begin{bmatrix}
1 & 0 & 0 & 0 \\
0 & 1 & 0 & 0 \\
0 & 0 & 0 & 1 \\
1 & 0 & 0 & 0 \\
0 & 1 & 0 & 0 \\
0 & 0 & 0 & 1 \\
1 & 0 & 0 & 0 \\
0 & 1 & 0 & 0 \\
0 & 0 & 0 & 1 \\
1 & 0 & 0 & 0 \\
0 & 1 & 1 & 0 \\
0 & 0 & 0 & 1 \\
1 & 0 & 0 & 0 \\
0 & 1 & 1 & 0 \\
0 & 0 & 0 & 1
\end{bmatrix}
\begin{bmatrix}
\beta_1 \\
\beta^2 \\
\beta_3 \\
\beta_4
\end{bmatrix}
$$

where β_1 represents the mean presidential trust for all respondents, β_2 represents the mean white Supreme Court trust, β_3 represents the black effect for the Supreme Court, and β_4 represents the mean Senate trust for all respondents. The chi-square goodness-of-fit statistic is 12.52 with 11 degrees of freedom, which is less than the critical value of $\chi^2_{11,\,0.75} = 13.70$. The estimated parameters are $b_1 = 0.44$, $b_2 = 0.17$, $b_3 = 0.26$, and $b_4 = 0.25$.

The parameter estimates from this final model suggest that the President enjoys the highest degree of trust with a mean trust level of 0.44 (44 percent), which implies that 44 percent more of the sample agree with the trust question than disagree. Average trust in the Senate is substantially lower ($b_4 = 0.25$). The Supreme Court presents a somewhat more complicated picture. Trust for the court among whites is relatively low ($b_2 = 0.17$); blacks are relatively more trusting than whites of this institution, since their mean trust score is 0.43 ($b_2 + b_3 = 0.17 + 0.26 = 0.43$). The level of trust exhibited by blacks for the Supreme Court is almost equal to the trust expressed by all citizens for the President. Table 8.3 contains the predicted trust scores based on model $X_2\beta_2$.

The final model $X_2\beta_2$ represents a parsimonious simplification of the original data: It requires only four parameters to estimate 15 functions. There is evidence of differing levels of trust for governmental institutions and the interaction of race with institution—specifically, blacks feel more positively about the Supreme Court than other citizens. We cannot say that this model is the true one, because we were unable to incorporate the complexities of the sample design into the estimation of V_F. Furthermore, we derived model $X_2\beta_2$ after discovering that the initial model $X_1\beta_1$ did not provide a parsimonious fit. Confirmation of these results will depend on an independent verification from another survey.

MULTIPLE PROPORTIONS

The mean trust function, though a useful summary measure of the marginal trust distributions, has several limitations that in some situations may make it unsuited for summarizing response variables with three or more levels. As previously indicated, the mean score function requires assumptions regarding the order of the categories and the distances between them. In this case the single mean function also may mask variation of interest to the researcher, because the not sure or no opinion category is weighted by zero. If the investigator is interested in the proportion of people in this category, the mean score provides no information about it.

One alternative is to summarize any k-level response variable with $(k - 1)$ multiple proportions. The use of multiple proportions is an effective way of studying all sources of variation in a categorical variable. We will demonstrate a multiple proportion approach by reanalyzing the trust data. Because we are conducting an after-the-fact analysis, our philosophical approach will be to account for the principal sources of variation in the data by testing groups of conceptually similar parameters—the main effects for race and income and the interactions among them. We will not focus on question differences directly, since we have prior knowledge that the institution affects the degree of trusting response. Our goal is to find a plausible model for future evaluation rather than to test each degree of freedom (source of variation) in the problem.

We define two proportions for each of the three trust questions within the five subpopulations. This approach requires defining $2 \times 3 \times 5 = 30$ functions. We have arbitrarily chosen to model the two proportions agreeing and disagreeing (and excluded the proportion not sure/no opinion).

The vector of observed functions **F** contains the probability of disagreement and the probability of agreement for the trust question about the President followed by the probabilities for the Supreme Court and Senate. It is formed as follows:

$$
\mathbf{F}_{(30 \times 135)} =
\begin{bmatrix}
\mathbf{A}^* & & & & \\
& \mathbf{A}^* & & & \\
& & \mathbf{A}^* & & \\
& & & \mathbf{A}^* & \\
& & & & \mathbf{A}^*
\end{bmatrix} \mathbf{p}
$$

where

$$
\mathbf{A}^* =
\begin{bmatrix}
111 & 111 & 111 & 000 & 000 & 000 & 000 & 000 & 000 \\
000 & 000 & 000 & 111 & 111 & 111 & 000 & 000 & 000 \\
111 & 000 & 000 & 111 & 000 & 000 & 111 & 000 & 000 \\
000 & 111 & 000 & 000 & 111 & 000 & 000 & 111 & 000 \\
100 & 100 & 100 & 100 & 100 & 100 & 100 & 100 & 100 \\
010 & 010 & 010 & 010 & 010 & 010 & 010 & 010 & 010
\end{bmatrix}
$$

The observed proportions are presented later in Table 8.5. The proportion agreeing exceeds the proportion disagreeing for all groups and all questions, and the President appears to evoke the highest levels of trusting responses. Trust in the Supreme Court and Senate varies by subpopulation. Note that by subtracting the proportion disagreeing from the proportion agreeing in Table 8.7 we obtain the observed mean trust scores in Table 8.3.

Table 8.5   Observed and Predicted Proportions Disagreeing and Agreeing to Three Trust Questions by Race and Income

Race and Income	Question	OBSERVED PROPORTIONS		PREDICTED PROPORTIONS[a]	
		Disagree	Agree	Disagree	Agree
White					
Low	President	0.19	0.62	0.20	0.65
	Court	0.30	0.50	0.34	0.51
	Senate	0.29	0.52	0.31	0.54
Medium	President	0.22	0.65	0.20	0.65
	Court	0.37	0.49	0.34	0.51
	Senate	0.37	0.53	0.31	0.54
High	President	0.21	0.70	0.20	0.71
	Court	0.34	0.52	0.34	0.51
	Senate	0.32	0.58	0.31	0.59
Black					
Low	President	0.17	0.61	0.20	0.56
	Court	0.17	0.63	0.19	0.60
	Senate	0.22	0.60	0.26	0.54
Medium	President	0.21	0.53	0.20	0.56
	Court	0.18	0.61	0.19	0.60
	Senate	0.27	0.49	0.26	0.54

[a] Based on model $X_5 \beta_5$.

We again wish to determine whether there are relationships between these probabilities and the race and income variables. To examine this question we use the model $X_3 \beta_3$, which contains a mean, the black effect, the medium-income effect, and the white and high-income effect (4 terms) for the probability of disagree and agree for each. Since there are two functions per question and three questions, we produce the following 24-DF ($4 \times 2 \times 3$) model:

$$\mathbf{X}_3\boldsymbol{\beta}_3 = \begin{bmatrix} 1000 & 0000 & 0000 & 0000 & 0000 & 0000 \\ 0000 & 1000 & & & & \\ 0000 & & 1000 & & & \\ 0000 & & & 1000 & & \\ 0000 & & & & 1000 & \\ 0000 & & & & & 1000 \\ 1010 & & & & & \\ 0000 & 1010 & & & & \\ 0000 & & 1010 & & & \\ 0000 & & & 1010 & & \\ 0000 & & & & 1010 & \\ 0000 & & & & & 1010 \\ 1001 & & & & & \\ 0000 & 1001 & & & & \\ 0000 & & 1001 & & & \\ 0000 & & & 1001 & & \\ 0000 & & & & 1001 & \\ 0000 & & & & & 1001 \\ 1100 & & & & & \\ 0000 & 1100 & & & & \\ 0000 & & 1100 & & & \\ 0000 & & & 1100 & & \\ 0000 & & & & 1100 & \\ 0000 & & & & & 1100 \\ 1110 & & & & & \\ 0000 & 1110 & & & & \\ 0000 & & 1110 & & & \\ 0000 & & & 1110 & & \\ 0000 & & & & 1110 & \\ 0000 & & & & & 1110 \end{bmatrix} \begin{bmatrix} \beta_1 \\ \beta_2 \\ \beta_3 \\ \beta_4 \\ \beta_5 \\ \beta_6 \\ \beta_7 \\ \beta_8 \\ \beta_9 \\ \beta_{10} \\ \beta_{11} \\ \beta_{12} \\ \beta_{13} \\ \beta_{14} \\ \beta_{15} \\ \beta_{16} \\ \beta_{17} \\ \beta_{18} \\ \beta_{19} \\ \beta_{20} \\ \beta_{21} \\ \beta_{22} \\ \beta_{23} \\ \beta_{24} \end{bmatrix}$$

We have omitted some of the zeros from the **X** matrix to highlight the pattern of 1's and 0's in the model. The 24 parameters are summarized in the following table:

	Trust for Low-Income Whites	Black Effects	Medium-Income Effects	White and High-Income Effects
Proportion disagree	$\beta_1, \beta_9, \beta_{17}$	$\beta_2, \beta_{10}, \beta_{18}$	$\beta_3, \beta_{11}, \beta_{19}$	$\beta_4, \beta_{12}, \beta_{20}$
Proportion agree	$\beta_5, \beta_{13}, \beta_{21}$	$\beta_6, \beta_{14}, \beta_{22}$	$\beta_7, \beta_{15}, \beta_{23}$	$\beta_8, \beta_{16}, \beta_{24}$

The three parameters listed in each cell of this summary table represent the effect for the President, Supreme Court, and Senate respectively. For example, β_7, β_{15}, and β_{23} are the parameters characterizing the medium-income effects on the proportion agreeing for the President, Supreme Court, and Senate respectively. Model $\mathbf{X}_3\boldsymbol{\beta}_3$ fits well since its $X_{\mathrm{GOF}}^2 = 7.11$ with 6 degrees of freedom does not exceed $\chi_{6,\,0.75}^2 = 7.84$.

We conducted six 3-DF tests corresponding to the six sets of three parameters grouped in the summary table above. We did not test the sets of mean parameters for low-income whites against zero, since this is our baseline group. The six tests are summarized here:

	Black Effects	Medium-Income Effects	White and High-Income Effects
Proportion disagree	$X^2 = 40.01$	$X^2 = 3.31$	$X^2 = 8.92$
Proportion agree	$X^2 = 17.03$	$X^2 = 2.05$	$X^2 = 6.60$

Critical value: $\chi_{3,\,0.99}^2 = 11.34$.

These six tests indicate that medium-income effects are not significant, and white and high-income effects are marginally significant. Black effects, however, play an important role in determining the respondent's answer to the trust questions.

We then fit an 18-DF model $\mathbf{X}_4\boldsymbol{\beta}_4$ (not shown) containing six low or medium income and white terms, six black effects, and six white and high income effects. The model's goodness-of-fit chi-square statistic is 2.22 with 12 degrees of freedom. Tests for the black and white and high-income effects are reported in Table 8.6. These tests show that four white and high-income effects and two black effects have observed X^2 values less than 1.00.

**Table 8.6   Selected Test Statistics $\mathbf{X}^2$
for Hypotheses Under Model $\mathbf{X}_4\boldsymbol{\beta}_4$**

	PROPORTION	
Description	*Disagree*	*Agree*
President		
Black effect	0.76	3.53
White and high-income effect	0.07	3.45
Senate		
Black effect	35.04	13.89
White and high-income effect	<0.01	0.69
Supreme Court		
Black effect	5.81	0.97
White and high-income effect	0.07	2.98

In seeking a parsimonious reduced model, we eliminate effects with observed chi-square values less than 1.00 since these effects probably contribute little to explaining variation in trust attitudes. This strategy is not theoretically satisfying, because it relies entirely on the results of the previous testing and is not guided by hypotheses formed prior to data analysis. Nevertheless, this approach will identify the principal sources of variation in the data set, and from this knowledge we may establish a model for further testing with a new data set.

By eliminating from model $\mathbf{X}_4 \boldsymbol{\beta}_5$ the six terms with observed X^2 less than 1.00, we obtain a 12-DF reduced model $\mathbf{X}_5 \boldsymbol{\beta}_5$:

$$
\mathbf{F} =
\begin{bmatrix}
1 & 000 & 00 & 00 & 00 & 00 \\
0 & 100 &   &   &   &   \\
0 &   & 10 &   &   &   \\
0 &   &   & 10 &   &   \\
0 &   &   &   & 10 &   \\
0 &   &   &   &   & 10 \\
1 &   &   &   &   &   \\
0 & 100 &   &   &   &   \\
0 &   & 10 &   &   &   \\
0 &   &   & 10 &   &   \\
0 &   &   &   & 10 &   \\
0 &   &   &   &   & 10 \\
1 &   &   &   &   &   \\
0 & 101 &   &   &   &   \\
0 &   & 10 &   &   &   \\
0 &   &   & 10 &   &   \\
0 &   &   &   & 10 &   \\
0 &   &   &   &   & 11 \\
1 &   &   &   &   &   \\
0 & 110 &   &   &   &   \\
0 &   & 11 &   &   &   \\
0 &   &   & 11 &   &   \\
0 &   &   &   & 11 &   \\
0 &   &   &   &   & 10 \\
1 &   &   &   &   &   \\
0 & 110 &   &   &   &   \\
0 &   & 11 &   &   &   \\
0 &   &   & 11 &   &   \\
0 &   &   &   & 11 &   \\
0 &   &   &   &   & 10
\end{bmatrix}
\begin{bmatrix}
\beta_1 \\ \beta_2 \\ \beta_3 \\ \beta_4 \\ \beta_5 \\ \beta_6 \\ \beta_7 \\ \beta_8 \\ \beta_9 \\ \beta_{10} \\ \beta_{11} \\ \beta_{12}
\end{bmatrix}
$$

The parameters are arranged as follows:

	Trust for Low or Medium Income and White	Black Effects	White and High-Income Effects
Proportion disagree	$\beta_1, \beta_5, \beta_9$	β_6, β_{10}	—
Proportion agree	$\beta_2, \beta_7, \beta_{11}$	β_3, β_8	β_4, β_{12}

The goodness-of-fit chi-square for this reduced model is $X^2_{GOF} = 15.59$ with 18 degrees of freedom, which is not statistically significant at $\alpha = 0.25$ ($\chi^2_{18,0.75} = 21.60$).

Table 8.7 shows that the four black effects and two white and high-income effects are statistically significant ($\alpha = 0.01$). The major sources in variation in response to the trust questions are the differences among the question content, black effects, and the white and high-income effects. The simultaneous test of the black effects has $X^2 = 54.70$. The least source of variation results from the white and high-income effects ($X^2 = 14.31$; DF = 2).

Table 8.7   Selected Test Statistics for Hypotheses Under Model $X_5\beta_5$

Parameter	Description	X^2	DF	b_i
β_3	Black effect: agree: President	12.21	1	−0.0●
β_6	Black effect: disagree: Senate	42.84	1	−0.1●
β_8	Black effect: agree: Senate	12.13	1	0.0●
β_{10}	Black effect: disagree: Supreme Court	6.50	1	−0.0●
$\beta_3, \beta_6, \beta_8, \beta_{10}$	Simultaneous test that 4 black effects equal zero	54.70	4	—
β_4	White and high-income effect: agree: President	9.36	1	0.0●
β_{12}	White and high-income effect: agree: Supreme Court	7.63	1	0.0●
β_4, β_{12}	Simultaneous test that 2 white and high-income effects equal zero	14.31	2	—

This is a parsimonious model in the sense that 12 estimated parameters adequately predict 30 functions. It is not as satisfactory as the final model fitted to the mean score function, because two special white and high-income effects (β_4 and β_{12}) and four race effects ($\beta_6, \beta_{10}, \beta_3, \beta_8$) are necessary to produce an adequate fit. Since the model $\mathbf{X}_5\boldsymbol{\beta}_5$ contains two white and income effects derived solely from data analysis, we cannot say that this is the true model. We have not pursued this analysis further since a major objective has been accomplished—to identify the principal sources of variation in responses to the three trust questions. Model $\mathbf{X}_5\boldsymbol{\beta}_5$ suggests that the major source of variation in trust responses results from the respondent's race and the question's content.

The observed and expected proportions disagreeing and agreeing are given in Table 8.5. The expected values show that the pattern of proportion agreeing and disagreeing differs for each question. Low or medium-income whites have the same pattern of trust responses; high-income whites differ from other whites in their tendency to trust the President and Senate more (agree response). Income does not affect the patterns of the response for low and medium-income blacks, but their levels of trust differ from those of whites. Black respondents tend both to disagree and to agree less to the trust questions than whites, which indicates a higher level of not sure/no opinion response for the group. The sum of the estimated disagree and agree proportions varies between 0.76 and 0.80 among blacks depending on the question, which indicates that 20 to 24 percent expressed a not sure/no opinion response. In comparison, 15 percent of low and medium-income whites and only 9 to 15 percent of high-income whites gave not sure/no opinion responses. The use of multiple proportions to characterize all variation in the trust question identifies a source of variation—disagree/agree versus not sure/no opinion—that is not reflected in the previous mean score analysis.

SUMMARY

In this chapter we demonstrated one analysis of a two-factor, three-response trust problem using functions constructed on the marginals of the response variable subtables. This analysis illustrates that regardless of the choice of function—either a mean score or multiple proportions—similar analytic questions exist. These questions relate to differences among subpopulations (race and income) and the interaction between factor (race and income) and response (trust) variables. The example in this chapter presents some complex interaction effects between the trust questions and the factor variables (race and income). Despite this complexity of the data, we were able to identify a parsimonious model that accounted for the significant sources of variation. Because of the after-the-fact nature of the analysis, we must view the model as suggestive of the true underlying relationship, confirmation of which will depend on data collected independently.

The key points covered in this chapter are:

- *The problem:* Compare levels of trust in the President, Supreme Court, and Senate by race and income categories of respondents.

- *Multiresponse table:* Layout can be similar to single-response analysis but the π_{ij} sum to 1 across the *combinations* of response within each subpopulation.
- *Mean trust scores:* The President has the highest level of trust; blacks also trust the Supreme Court highly, while whites have a much lower level of trust in it.
- *Multiple proportions:* Blacks have a higher proportion of not sure/no opinion responses; high-income whites have more trust in the President and Senate than other whites; blacks have less trust in the President and Senate but more trust in the Supreme Court.

This chapter has provided two ways of examining multiple responses. The mean score approach gives an overall view of the central tendency, but it can obscure other important information in the data. The multiple proportion analysis, though more complicated, revealed that blacks had higher proportions of not sure/no opinion responses than whites—a fact hidden by the use of the mean trust score. In the next chapter we present another way of studying multiple responses, one based on a rank correlation method. This method is of special interest to analysts familiar with applications of the Pearson product-moment correlation coefficient in parametric analyses.

EXERCISES

8.1.

The data in Table 8.8 represent the agree/disagree responses to two administrations of the statement "The courts are a major cause of breakdown of law and order" by the respondent's level of education.

    a. Construct the marginal functions of the probabilities of disagreeing and agreeing for each administration within each of the three subtables. This task requires four functions for each of the three subtables—that is, 12 functions.

    b. Test the hypotheses of *marginal homogeneity* for each level of education:

$$H_0: \begin{cases} \pi_1(\text{disagree}) = \pi_2(\text{disagree}) & 1 = \text{time 1} \\ \pi_1(\text{agree}) = \pi_2(\text{agree}) & 2 = \text{time 2} \end{cases}$$

    c. Fit a model estimating education effects on marginal homogeneity. Fit a reduced model that adequately characterizes variation in the distribution of response to each administration of the question by education. Write a brief paragraph interpreting your findings.

**Table 8.8 Responses of 693 People to Two
Administrations of a Question by Education**

FIRST RESPONSE	SECOND RESPONSE		
	Disagree	*Not Sure*	*Agree*
Less than high school			
Disagree	30	5	12
Not sure	5	9	8
Agree	12	8	92
High school			
Disagree	31	5	21
Not sure	8	13	7
Agree	14	6	92
Some college			
Disagree	66	9	30
Not sure	9	18	8
Agree	28	13	134

Source: Houston Community Study data files; Stafford and
Lehnen (1975).

8.2.

The analysis of marginal functions may be used to test net changes across time.
Use the 1948 presidential election study panel data presented in Exercise 7.4
(Chapter 7) to analyze the net change in attitudes toward the Republican candi-
date and voting preferences across two time periods.

a. Form the following functions:

$\pi_{R,1}$ = probability preferring Republican candidate at time 1

$\pi_{+,1}$ = probability having favorable view of Republican candidate at
time 1

$\pi_{R,2}$ = probability preferring Republican candidate at time 2

$\pi_{+,2}$ = probability having favorable view of Republican candidate at
time 2

b. Conduct the following hypothesis tests:

1. H_0: $\pi_{R,1} - \pi_{R,2} = 0$
2. H_0: $\pi_{+,1} - \pi_{+,2} = 0$

c. Answer the following questions:

1. Did the level of support for the Republican candidate change from time 1 to time 2? Describe the change.
2. Did the level of favorable attitude toward the Republican candidate change from time 1 to time 2? Describe the change.

8.3.

Use the British-Danish intergenerational mobility data presented in Exercise 6.1 (Chapter 6) and do the following analyses.

a. Construct a mean status score for *British* fathers ($\bar{X}_{BF}$) and sons ($\bar{X}_{BS}$) and test whether the population mean scores are equal.

b. Perform an analysis identical to part (a) on the *Danish* father–son data. Is the average status of fathers and sons the same in each country?

c. Analyze the British-Danish data as one table and answer the following questions. *Hint:* The frequency table for this analysis is of size (2×25).

1. Are the mean status scores for British fathers and sons equal?
2. Are the mean status scores for Danish fathers and sons equal?
3. Is there a country effect on the status of fathers and sons?
4. Are the effects of each country on intergenerational mobility the same?

d. In what ways is the analysis performed in part (c) similar to and different from the analyses performed in parts (a) and (b)?

e. Analyze the British data using the functions

$$\mathbf{F} = \begin{bmatrix} p_{F1} \\ p_{F2} \\ p_{F3} \\ p_{F4} \\ p_{S1} \\ p_{S2} \\ p_{S3} \\ p_{S4} \end{bmatrix}$$

where p_{ij} is a marginal proportion; the subscripts F = father and S = son; and 1, 2, 3, and 4 represent the status level. Thus the proportion p_{S3} is the proportion of sons at occupational status level 3.

1. Construct a 4-DF test to determine whether the distributions of father and son status are equal.

2. Test the equality of father and son status for each level.

3. Test the equality of father and son status for level 5.

4. Summarize your findings in a paragraph.

   f. This exercise has suggested several approaches to analyzing the British-Danish intergenerational mobility data. Which method is best? Explain your answer. Suggest alternative approaches for analyzing these data.

APPENDIX: GENCAT INPUT

The following listing is for the GENCAT control cards that reproduce the analysis described in this chapter.

```
    5     2     1                    CHAPTER 8: ANALYSIS OF TRUST DATA
    5    27
    3                                      (40F2.0)
-1-1-1-1-1-1-1-1-1010101010101010101
-1-1-1010101000000-1-1-1010101000000-1-1-1010101
-10100-10100-10100-10100-10100-10100-10100-10100-101
       15.          3.          1.          5.          8.          4.          4.          2.
        3.         13.         17.          5.         12.         60.         10.          4.
       12.          8.          9.          3.          2.          2.         10.          2.
        2.          3.          7.
       41.         17.          4.         12.         17.          0.          2.          2.
        1.         37.         31.         10.         29.        127.         15.          8.
       17.         13.         12.          6.          5.          2.         10.          5.
        2.          6.         12.
       35.         10.          4.         21.         12.          0.          2.          1.
        3.         30.         46.          2.         18.        136.         12.          8.
       24.         10.         11.          2.          1.          4.          6.          5.
        3.          1.          3.
       12.          4.          2.          0.          6.          0.          4.          1.
        3.          8.          1.          1.          3.         81.          7.          3.
        6.          6.          2.          2.          0.          7.         10.          6.
        3.          3.          8.
        8.          1.          1.          4.          9.          1.          0.          0.
        1.          3.          2.          0.         10.         30.          8.          2.
        7.          1.          3.          2.          2.          0.          6.          5.
        2.          1.         10.
    7     1    12                    (80F1.0)          INITIAL MODEL
100100100100100
000000000100100
000100000000100
000000100000000
010010010010010
000000000010010
000010000000010
000000010000000
001001001001001
000000000001001
000001000000001
000000001000000
    8     1     3                    (40F2.0)          RACE EFFECTS
  1
             1
                         1
    8     1     1                    (40F2.0)
  1
    8     1     1                    (40F2.0)
             1
    8     1     1                    (40F2.0)
                         1
    8     1     3                    (40F2.0)          HI & WHITE EFFECTS
      1
```

```
                      1
                            1
      8     1     1              (40F2.0)
         1
      8     1     1              (40F2.0)
                         1
      8     1     1              (40F2.0)
                               1
      8     1     3              (40F2.0)         MEDIUM INCOME EFFECTS
         1
                     1
                                 1
      8     1     1              (40F2.0)
         1
      8     1     1              (40F2.0)
                     1
      8     1     1              (40F2.0)
                           1
      8     1     8              (40F2.0)
   1
                        1
      1
                  1
                        1
   1
               1
                         1
      7     1     4              (80F1.0)       REDUCED   MODEL
100100100100100
010010010010010
000000000010010
001001001001001
      5     2     1                      PART 2:ANALYSIS OF TRUST DATA
      5    27
      6                          (80F1.0)
111111111
000000000111111111
111000000111000000111
000111000000111000000111
100100100100100100100001
010010010010010010010001
```

15.	3.	1.	5.	8.	4.	4.	2.
3.	13.	17.	5.	12.	60.	10.	4.
12.	8.	9.	3.	2.	2.	10.	2.
2.	3.	7.					
41.	17.	4.	12.	17.	0.	2.	2.
1.	37.	31.	10.	29.	127.	15.	8.
17.	13.	12.	6.	5.	2.	10.	5.
2.	6.	12.					
35.	10.	4.	21.	12.	0.	2.	1.
3.	30.	46.	2.	18.	136.	12.	8.
24.	10.	11.	2.	1.	4.	6.	5.
3.	1.	3.					
12.	4.	2.	0.	6.	0.	4.	1.
3.	8.	1.	1.	3.	81.	7.	3.
6.	6.	2.	2.	0.	7.	10.	6.
3.	3.	8.					
8.	1.	1.	4.	9.	1.	0.	0.
1.	3.	2.	0.	10.	30.	8.	2.
7.	1.	3.	2.	2.	0.	6.	5.
2.	1.	10.					

```
      7     1    24              (80F1.0)       INITIAL MODEL
100000100001000010000010000010000
000000000000000000100000100000
000001000000000000000001
000000000001000000000000
010000100001000010000010000
000000000000000000010000010
000000100000000000000001
000000000000010
001000001000001000001000001000
000000000000000000001000001
000000001000000000000000001
```

```
000000000000001
000100000100000100000100000100
000000000000000000000100000100
000000001000000000000000000001
000000000000000100
000010000010000010000010000010
000000000000000000000010000010
000000000010000000000000000010
00000000000000001
000001000001000001000001000001
000000000000000000000001000001
000000000010000000000000000001
000000000000000001
     8    1    3              (80F1.0)           TOTAL RACE EFFECTS/DISAGREE
01
000001
0000000001
     8    1    3              (80F1.0)           TOTAL RACE EFFECTS/AGREE
00000000000001
000000000000000001
000000000000000000001
     8    1    3              (80F1.0)           MED INCOME EFFECTS/DISAGREE
001
0000001
00000000001
     8    1    3              (80F1.0)           MED INCOME EFFECTS/AGREE
000000000000001
000000000000000001
000000000000000000001
     8    1    3              (80F1.0)           HI & WHITE/DISAGREE
0001
00000001
000000000001
     8    1    3              (80F1.0)           HI & WHITE/AGREE
000000000000001
000000000000000001
000000000000000000001
     7    1   18              (80F1.0)           REDUCED MODEL
100000100000100000100000100000
000000000000000000100000100000
000000000000100000000000
010000100000100000100000010000
000000000000000000100000010
000000000000010
001000001000001000001000001000
000000000000000000001000001
000000000000001
000100000100000100000100000100
000000000000000000000100000100
000000000000000100
000010000010000010000010000010
000000000000000000000010000010
000000000000000001
000001000001000001000001000001
000000000000000000000001000001
000000000000000001
     8    1    1              (80F1.0)
01
     8    1    1              (80F1.0)
001
     8    1    1              (80F1.0)
00001
     8    1    1              (80F1.0)
000001
     8    1    1              (80F1.0)
00000001
     8    1    1              (80F1.0)
000000001
     8    1    1              (80F1.0)
00000000001
     8    1    1              (80F1.0)
000000000001
     8    1    1              (80F1.0)
00000000000001
```

```
     8    1    1            (80F1.0)
000000000000001
     8    1    1            (80F1.0)
00000000000000001
     8    1    1            (80F1.0)
000000000000000001
     8    1    3            (80F1.0)              HI & WHITE/DISAGREE
001
000000001
000000000000001
     8    1    3            (80F1.0)              HI & WHITE/AGREE
000001
000000000001
000000000000000001
     8    1    6            (80F1.0)              TOTAL HIGH & WHITE EFFECTS
001
000001
000000001
000000000001
000000000000001
000000000000000001
     7    1    15           (80F1.0)
100000100000100000100000100000
00000000000000000000100000100000
01000001000001000001000001000001
00000000000000000000100000010
000000000000010
00100000100000100000100000110000
00000000000000000000100000100001
00010000010000010000010000010100
00000000000000000000010000010000010100
000000000000000100
00001000010000010000010000010
00000000000000000000100000010
00000100000100000100000100000100001
00000000000000000000000000010100001
000000000000000001
     8    1    3            (80F1.0)              HI & WHITE/AGREE
00001
0000000001
000000000000001
     8    1    1            (80F1.0)
00001
     8    1    1            (80F1.0)
0000000001
     8    1    1            (80F1.0)
000000000000001
     8    1    1            (40F2.0)
0000000001000000000000000000-1
     7    1    13           (80F1.0)
100000100000100000100000100000
0000000000000000000100000100000
01000001000001000001000001000001
00000000000000000000100000010
00100000100000100000010000001000
00000000000000000000100000100001
0001000001000001000001000001000100
00000000000000000000010000010000100
00001000010000010000010000010
00000000000000000000000010000010
00000100000100000100000100000100001
0000000000000000000000000010100001
000000000000010001
     8    1    1            (80F1.0)
01
     8    1    1            (80F1.0)
0001
     8    1    1            (80F1.0)
000001
     8    1    1            (80F1.0)
00000001
     8    1    1            (80F1.0)
0000000001
```

```
     8     1     1                (80F1.0)
000000000001
     8     1     1                (80F1.0)
0000000000001
     7     1    12                (80F1.0)
100000100000100000100000100000
000000000000000000100000100000
010000100000100000010000010000
000000000000000000010000010
001000001000001000001000001000
000000000000000000001000001
000100000100000100000100000100
000000000000000000000100000100
000010000010000010000010000010
000000000000000000000010000010
000001000001000001000001000001
000000000000000000000001000001
```

9
Rank Correlation Methods

Rank correlation coefficients are statistical indices that measure the degree of association between two variables having ordered categories. Some well-known rank correlation coefficients are those proposed by Goodman and Kruskal (1954, 1959), Kendall (1955), and Somers (1962). Rank correlation methods share several common features. They are based on counts and are defined such that a coefficient of zero means "no association" between the variables and a value of $+1.0$ or -1.0 means "perfect agreement" or "perfect inverse agreement," respectively.

Rank correlation methods have been popular in social science applications because of their analogy to Pearson product-moment correlation coefficients. They have had limited use in multivariate applications, however, because of the difficulty of computing tests of significance and using them for multivariate analysis. In the following pages we illustrate the use of rank correlation coefficients and demonstrate that these statistics are one example of a large class of complex functions that may be defined from the cell frequencies in the context of WLS theory.

ARE PEOPLE CONSISTENT IN THEIR RESPONSES?

Many urban areas in the United States face problems created by rapid growth. Houston, Texas, is one of the more extreme examples. In the middle 1970s

146

one study estimated that Houston was growing at the rate of 1000 new people each week—more than 50,000 each year (Stafford and Lehnen, 1975).

Such rapid growth creates major problems for the delivery of urban services and profoundly affects the city's quality of life. The capacity of organizations to provide education, sanitation, mass transportation, housing, and public health is often strained to the limit, if not overwhelmed, and the quality of life—measured by social indicators such as air quality and crime statistics— declines. Perceiving this all-too-familiar scenario, three Houston-based corporations commissioned a research project, the Houston Community Study, to measure among other things the attitudes and opinions of a cross-section of Houstonians toward the problems created by urban growth (Stafford and Lehnen, 1975). The data were to be used by business firms and social organizations based in the city for establishing priorities among community service projects.

The sponsors of the project were concerned that public attitudes toward policy issues could not be measured reliably. Previous opinion research conducted by the University of Michigan Survey Research Center reported that some sectors of the public lacked sufficient knowledge and attentiveness to form stable opinions about broad questions of policy (Converse, 1964). This conclusion was arrived at by comparing responses among a set of respondents to successive administrations of the same question. The varying patterns of consistency and inconsistency suggested that certain issues, especially questions of domestic policy, were beyond the attention range of respondents with lower social status. In response to these considerations, the study team developed a methodological study as part of a household sample survey to measure the consistency of response among various groups of individuals and to test the effect of education on response consistency.

Fifteen sets of policy statements were administered by means of a card-sort technique. At two points during the interview respondents were given a deck of about 50 cards containing statements eliciting an agree/disagree response. Respondents were asked to sort the cards into boxes labeled "agree," "disagree," "have no opinion," and "not sure." Since identical repetitions of policy statements were included in each card deck, the respondents replied twice to the same statement during the course of the interview.

Table 9.1 reports response data for two administrations of the statement measuring attitudes toward integration of the schools by level of the respondent's education. Each subtable is size (4×4), representing the 16 combinations of two responses to the question "The government should make sure that white and Negro children go to the same school together."

Why is response consistency so low? One prevailing theory argues that respondents, desiring to "please," to "be helpful," or to hide their lack of understanding, respond to a question even when they have no idea of its meaning. To reduce this bias, interviewers instructed the respondents in the difference

Table 9.1   Responses to the Statement "The Government Should Make Sure That White and Negro Children Go to the Same School Together"

(a) Respondents with Less Than High School Education (n = 252)

	SECOND RESPONSE			
FIRST RESPONSE	Disagree	Not Sure	Agree	Have No Opinion
Disagree	86	4	13	3
Not Sure	5	5	8	3
Agree	11	5	74	4
Have no opinion	7	2	7	15

(b) Respondents with High School Diploma (n = 235)

	SECOND RESPONSE			
FIRST RESPONSE	Disagree	Not Sure	Agree	Have No Opinion
Disagree	86	5	13	2
Not Sure	5	0	8	2
Agree	16	1	61	6
Have no opinion	4	4	12	10

(c) Respondents with Some College Education (n = 335)

	SECOND RESPONSE			
FIRST RESPONSE	Disagree	Not Sure	Agree	Have No Opinion
Disagree	144	5	25	4
Not Sure	7	23	10	1
Agree	14	4	67	5
Have no opinion	5	4	3	14

Table 9.2   Presence or Absence of Opinion by Education

(a) Less Than High School ($n = 252$)			(b) High School ($n = 235$)			(c) Some College ($n = 335$)		
	0	*No*		*0*	*No*		*0*	*No*
0	211	10	0	195	10	0	299	10
No	16	15	No	20	10	No	12	14

0 = expressed agree/not sure/disagree response; No = "have no opinion" response.

between "not having an opinion" and being ambivalent ("not sure"). Thus the "not sure" box was placed in the middle of the agree/disagree continuum and the "have no opinion" box was placed to one side of the card-sort board. The data arrayed in Table 9.1 reflect this distinction; the respondents who indicated opinions on two occasions are grouped in the 3 × 3 area in the upper left of each subtable.

Two different patterns of consistency are of interest here. One is for the subset of each education level who expressed agree/not sure/disagree opinions twice. The second pattern analyzed is the opinion versus "have no opinion" comparison. Each subtable in Table 9.1 can be reduced to 2 × 2 tables as shown in Table 9.2. These three subtables are used to analyze one function; the 3 × 3 subtables for people expressing an opinion twice are used to construct a second function.

A RANK CORRELATION COEFFICIENT

Although data such as those reported in Table 9.1 present many alternatives for statistical analysis, we will demonstrate the approach common to the sociological literature (Converse, 1964). This approach uses rank correlation measures, such as Goodman and Kruskal's gamma, to measure the degree of consistency between two responses. The *gamma statistic* is a convenient choice because it has a possible range of -1.0 to $+1.0$, and it may be interpreted as the *proportion of untied pairs* of respondents that *agree* minus the proportion that *disagree*. Gamma measures only one type of response consistency: the tendency of the responses to group along the major or minor diagonals of a table.

The following two examples demonstrate that gamma varies according to the distribution of the data in the body of the table even though the marginal distributions remain the same. Note that the body of the tables shows considerable differences in *gross* change, to which the gamma statistics are sensitive, although there are no differences in the marginal distributions:

Example of Frequency Table
Having Low Gamma ($\hat{G} = -0.17$)

RESPONSE TO FIRST QUESTION	RESPONSE TO SECOND QUESTION		
	Agree	*Disagree*	*Total*
Agree	20	28	48
Disagree	20	20	40
Total	40	48	88

Example of Frequency Table
Having High Negative Gamma ($\hat{G} = -0.98$)

RESPONSE TO FIRST QUESTION	RESPONSE TO SECOND QUESTION		
	Agree	*Disagree*	*Total*
Agree	4	44	48
Disagree	36	4	40
Total	40	48	88

Rank correlation coefficients such as gamma, Kendall's tau, and Somers' d_{yx} and d_{xy} are some of the more complicated functions that may be analyzed in the context of the WLS approach (Forthofer and Koch, 1973). Rank correlation coefficients require the computation of the following quanitities:

$$A = \text{number of pairs in agreement}$$

$$D = \text{number of pairs in disagreement}$$

$$T_x = \text{number of pairs tied on variable } X$$

$$T_y = \text{number of pairs tied on variable } Y$$

$$T_{xy} = \text{number of pairs tied on both variables}$$

$$n = \text{sample size}$$

$$\text{TOT} = \text{total number of pairs} = A + D + T_x + T_y + T_{xy} = n(n - 1)/2$$

The estimates of Goodman and Kruskal's gamma (G), Kendall's tau (T), and Somers' measures (d_{yx} and d_{xy}) use the difference $A - D$ in the numerator. The estimate of gamma (G) is defined as

$$G = \frac{A - D}{A + D}$$

where the denominator is the number of untied pairs only. Kendall's tau is estimated by

$$T = \frac{A - D}{\sqrt{(A + D + T_x)(A + D + T_y)}}$$

Estimates of Somers' asymmetric measures d_{yx} and d_{xy} are defined as

$$d_{yx} = \frac{A - D}{A + D + T_y}$$

$$d_{xy} = \frac{A - D}{A + D + T_x}$$

In general, the various rank correlation coefficients differ in their treatment of the tied pair counts T_x, T_y, and T_{xy} in the denominator. Each function may be analyzed by using the WLS approach.

Table 9.3 summarizes the algebraic computations for rank correlations for the agree/not sure/disagree and the opinion/no opinion patterns for data taken from panel (a) of Table 9.1. The example illustrates that all operations are based on the 16 cell counts $n_{11}, n_{12}, \ldots, n_{1,16}$. We obtain the count of concordant pairs (A) by starting with the upper left-hand cell frequency and multiplying it by every cell count that is below and to the right—that is, $n_{11}(n_{16} + n_{17} + n_{1,10} + n_{1,11})$. The cells below and to the right have more positive feelings (not sure and agree are more positive than disagree) on *both* administrations of the question than the cell that contains people who said disagree both times. We take every cell in the table with observations below and to the right—namely, cells n_{11}, n_{12}, n_{15}, and n_{16}—and perform a similar operation. Summing these counts produces A.

To determine the number of discordant pairs (D), we use a similar rule except that we start in the upper right-hand corner and add the products of cells below and to the left. Anderson and Zelditch (1968) provide an easy-to-read description of the computational steps for deriving the counts for A, D, T_x, T_y, T_{xy}, and TOT and calculating the values of several rank correlation coefficients.

Table 9.3   Computing Gamma Coefficients for Data Reported in Table 9.1 (Panel a)

(a) Frequencies n_{ij}

Variable 1	Variable 2 Low ← High →		No Opinion
Low	$n_{11} = 86$	$n_{12} = 4$ $n_{13} = 13$	$n_{14} = 3$
↕	$n_{15} = 5$	$n_{16} = 5$ $n_{17} = 8$	$n_{18} = 3$
High	$n_{19} = 11$	$n_{1,10} = 5$ $n_{1,11} = 74$	$n_{1,12} = 4$
No opinion	$n_{1,13} = 7$	$n_{1,14} = 2$ $n_{1,15} = 7$	$n_{1,16} = 15$

(b) Computing Gamma for Consistency of Response (among Ordered Categories)

Number of Pairs in Agreement (A)

$n_{11}(n_{16} + n_{17} + n_{1,10} + n_{1,11}) = 86(5 + 8 + 5 + 74) = 7912$
$n_{12}(n_{17} + n_{1,11}) = 4(8 + 74) = 328$
$n_{15}(n_{1,10} + n_{1,11}) = 5(5 + 74) = 395$
$n_{16}(n_{1,11}) = 5(74) = \underline{370}$
$A = 9005$

Number of Pairs in Disagreement (D)

$n_{13}(n_{15} + n_{16} + n_{19} + n_{1,10}) = 13(5 + 5 + 11 + 5) = 338$
$n_{12}(n_{15} + n_{19}) = 4(5 + 11) = 64$
$n_{17}(n_{19} + n_{1,10}) = 8(11 + 5) = 128$
$n_{16}(n_{19}) = 5(11) = \underline{55}$
$D = 585$

$$\text{Gamma}(G) = \frac{A - D}{A + D} = \frac{9005 - 585}{9005 + 585} = 0.88.$$

(c) Computing Gamma for Consistency of Response (Opinion vs. No Opinion)

Number of Pairs in Agreement (A)

$(n_{11} + n_{12} + n_{13} + n_{15} + n_{16} + n_{17} + n_{19} + n_{1,10} + n_{1,11})n_{1,16} = 211(15) = \underline{3165}$
$A = 3165$

Number of Pairs in Disagreement (D)

$(n_{14} + n_{18} + n_{1,12})(n_{1,13} + n_{1,14} + n_{1,15}) = 10(16) = \underline{160}$
$D = 160$

$$\text{Gamma}(G) = \frac{A - D}{A + D} = \frac{3165 - 160}{3165 + 160} = 0.90.$$

COMPUTING GAMMA USING GENCAT

We demonstrate the creation of the two gammas shown in Table 9.3 by using the matrix operations found in the GENCAT program. Although there are more efficient methods for computing the correlation coefficients and their variance-covariance estimates, this example shows how a complex function is produced by a combination of linear, logarithmic, and exponential transformation of the cell counts.

The data for a single population are arranged as in Table 9.1. The vector of observed probabilities is

$$\mathbf{p}_1 \atop (16 \times 1) = \begin{bmatrix} p_{11} \\ p_{12} \\ \vdots \\ p_{1,16} \end{bmatrix}$$

The calculations shown in Table 9.3 indicate that the following 20 quantities are required to find the number of concordant and discordant pairs:

Concordant	*Discordant*
Opinion vs. No Opinion	Opinion vs. No Opinion
$(n_{11}+n_{12}+n_{13}+n_{15}+n_{16}+n_{17}+n_{19}+n_{1,10}+n_{1,11})$	$(n_{14}+n_{18}+n_{1,12})$
$n_{1,16}$	$(n_{1,13}+n_{1,14}+n_{1,15})$
Agree/Disagree	Agree/Disagree
n_{11}	n_{13}
$(n_{16}+n_{17}+n_{1,10}+n_{1,11})$	$(n_{15}+n_{16}+n_{19}+n_{1,10})$
n_{12}	n_{12}
$(n_{17}+n_{1,11})$	$(n_{15}+n_{19})$
n_{15}	n_{17}
$(n_{1,10}+n_{1,11})$	$(n_{19}+n_{1,10})$
n_{16}	n_{16}
$n_{1,11}$	n_{19}

Each of these 20 quantities is an individual cell count or the sum of the counts. We can create the sums and quantities listed above by forming linear combinations of the probabilities—that is, by premultiplying $\mathbf{p}_1$ by the matrix $\mathbf{A}_1$:

$$\mathbf{A}_1 \atop (20 \times 16) = \begin{bmatrix} \mathbf{A}_{11} \\ (2 \times 16) \\ \mathbf{A}_{12} \\ (2 \times 16) \\ \mathbf{A}_{13} \\ (8 \times 16) \\ \mathbf{A}_{14} \\ (8 \times 16) \end{bmatrix}$$

where

$$\text{Opinion vs. No Opinion, Concordant}$$

$$\mathbf{A}_{11}_{(2\times16)} = \begin{bmatrix} 1 & 1 & 1 & 0 & 1 & 1 & 1 & 0 & 1 & 1 & 1 & 0 & 0 & 0 & 0 & 0 \\ 0 & 0 & 0 & 0 & 0 & 0 & 0 & 0 & 0 & 0 & 0 & 0 & 0 & 0 & 0 & 1 \end{bmatrix}$$

$$\text{Opinion vs. No Opinion, Discordant}$$

$$\mathbf{A}_{12}_{(2\times16)} = \begin{bmatrix} 0 & 0 & 0 & 1 & 0 & 0 & 0 & 1 & 0 & 0 & 0 & 1 & 0 & 0 & 0 & 0 \\ 0 & 0 & 0 & 0 & 0 & 0 & 0 & 0 & 0 & 0 & 0 & 0 & 1 & 1 & 1 & 0 \end{bmatrix}$$

$$\text{Agree/Disagree, Concordant}$$

$$\mathbf{A}_{13}_{(8\times16)} = \begin{bmatrix} 1 & 0 & 0 & 0 & 0 & 0 & 0 & 0 & 0 & 0 & 0 & 0 & 0 & 0 & 0 & 0 \\ 0 & 0 & 0 & 0 & 0 & 1 & 1 & 0 & 0 & 1 & 1 & 0 & 0 & 0 & 0 & 0 \\ 0 & 1 & 0 & 0 & 0 & 0 & 0 & 0 & 0 & 0 & 0 & 0 & 0 & 0 & 0 & 0 \\ 0 & 0 & 0 & 0 & 0 & 0 & 1 & 0 & 0 & 0 & 1 & 0 & 0 & 0 & 0 & 0 \\ 0 & 0 & 0 & 0 & 1 & 0 & 0 & 0 & 0 & 0 & 0 & 0 & 0 & 0 & 0 & 0 \\ 0 & 0 & 0 & 0 & 0 & 0 & 0 & 0 & 0 & 1 & 1 & 0 & 0 & 0 & 0 & 0 \\ 0 & 0 & 0 & 0 & 0 & 1 & 0 & 0 & 0 & 0 & 0 & 0 & 0 & 0 & 0 & 0 \\ 0 & 0 & 0 & 0 & 0 & 0 & 0 & 0 & 0 & 0 & 1 & 0 & 0 & 0 & 0 & 0 \end{bmatrix}$$

$$\text{Agree/Disagree, Discordant}$$

$$\mathbf{A}_{14}_{(8\times16)} = \begin{bmatrix} 0 & 0 & 1 & 0 & 0 & 0 & 0 & 0 & 0 & 0 & 0 & 0 & 0 & 0 & 0 & 0 \\ 0 & 0 & 0 & 0 & 1 & 1 & 0 & 0 & 1 & 1 & 0 & 0 & 0 & 0 & 0 & 0 \\ 0 & 1 & 0 & 0 & 0 & 0 & 0 & 0 & 0 & 0 & 0 & 0 & 0 & 0 & 0 & 0 \\ 0 & 0 & 0 & 0 & 1 & 0 & 0 & 0 & 1 & 0 & 0 & 0 & 0 & 0 & 0 & 0 \\ 0 & 0 & 0 & 0 & 0 & 0 & 1 & 0 & 0 & 0 & 0 & 0 & 0 & 0 & 0 & 0 \\ 0 & 0 & 0 & 0 & 0 & 0 & 0 & 0 & 1 & 1 & 0 & 0 & 0 & 0 & 0 & 0 \\ 0 & 0 & 0 & 0 & 0 & 1 & 0 & 0 & 0 & 0 & 0 & 0 & 0 & 0 & 0 & 0 \\ 0 & 0 & 0 & 0 & 0 & 0 & 0 & 0 & 1 & 0 & 0 & 0 & 0 & 0 & 0 & 0 \end{bmatrix}$$

The next step is to create the products indicated in Table 9.3. Since we know that $\ln(a) + \ln(b) = \ln(ab)$, we may add logarithms of the appropriate terms. Taking the logarithms of the 20 terms formed by $\mathbf{A}_1\mathbf{p}_1$ yields $\ln(\mathbf{A}_1\mathbf{p}_1)$; adding the appropriate pairs of logarithms produces the logarithms of the desired products. The $\mathbf{A}_2$ matrix that forms the sums of the appropriate pairs is

$$\mathbf{A}_2 \atop (10 \times 20) = \begin{bmatrix} 1 & 1 & 0 & 0 & 0 & 0 & 0 & 0 & 0 & 0 & 0 & 0 & 0 & 0 & 0 & 0 & 0 & 0 & 0 & 0 \\ 0 & 0 & 1 & 1 & 0 & 0 & 0 & 0 & 0 & 0 & 0 & 0 & 0 & 0 & 0 & 0 & 0 & 0 & 0 & 0 \\ 0 & 0 & 0 & 0 & 1 & 1 & 0 & 0 & 0 & 0 & 0 & 0 & 0 & 0 & 0 & 0 & 0 & 0 & 0 & 0 \\ 0 & 0 & 0 & 0 & 0 & 0 & 1 & 1 & 0 & 0 & 0 & 0 & 0 & 0 & 0 & 0 & 0 & 0 & 0 & 0 \\ 0 & 0 & 0 & 0 & 0 & 0 & 0 & 0 & 1 & 1 & 0 & 0 & 0 & 0 & 0 & 0 & 0 & 0 & 0 & 0 \\ 0 & 0 & 0 & 0 & 0 & 0 & 0 & 0 & 0 & 0 & 1 & 1 & 0 & 0 & 0 & 0 & 0 & 0 & 0 & 0 \\ 0 & 0 & 0 & 0 & 0 & 0 & 0 & 0 & 0 & 0 & 0 & 0 & 1 & 1 & 0 & 0 & 0 & 0 & 0 & 0 \\ 0 & 0 & 0 & 0 & 0 & 0 & 0 & 0 & 0 & 0 & 0 & 0 & 0 & 0 & 1 & 1 & 0 & 0 & 0 & 0 \\ 0 & 0 & 0 & 0 & 0 & 0 & 0 & 0 & 0 & 0 & 0 & 0 & 0 & 0 & 0 & 0 & 1 & 1 & 0 & 0 \\ 0 & 0 & 0 & 0 & 0 & 0 & 0 & 0 & 0 & 0 & 0 & 0 & 0 & 0 & 0 & 0 & 0 & 0 & 1 & 1 \end{bmatrix}$$

$$\mathbf{A}_2 \atop (10 \times 20) = \begin{bmatrix} \mathbf{A}_2^* \atop (1 \times 2) & & & \\ & \mathbf{A}_2^* & & \\ & & \ddots & \\ & & & \mathbf{A}_2^* \end{bmatrix}$$

where $\mathbf{A}_2^* = \begin{bmatrix} 1 & 1 \end{bmatrix}$. The diagonal structure of $\mathbf{A}_2$ results from the $\mathbf{A}_1$ matrix, which puts terms to be multiplied adjacent to one another.

If we take the exponential of the logarithmic functions, we now have the desired products. The next step is to form $A - D$ and $A + D$. This is done by summing the appropriate products as indicated in Table 9.3. The $\mathbf{A}_3$ matrix performs this summation:

row label

$$\mathbf{A}_3 \atop (4 \times 10) = \begin{bmatrix} 1 & -1 & 0 & 0 & 0 & 0 & 0 & 0 & 0 & 0 \\ 1 & 1 & 0 & 0 & 0 & 0 & 0 & 0 & 0 & 0 \\ 0 & 0 & 1 & 1 & 1 & 1 & -1 & -1 & -1 & -1 \\ 0 & 0 & 1 & 1 & 1 & 1 & 1 & 1 & 1 & 1 \end{bmatrix} \begin{array}{l} A - D \text{ (opinion/no opinion)} \\ A + D \text{ (opinion/no opinion)} \\ A - D \text{ (agree/disagree)} \\ A + D \text{ (agree/disagree)} \end{array}$$

We now have formed the numerator and denominators of the gammas. To take their ratio, we again use logarithms because $\ln a - \ln b = \ln(a/b)$. Premultiply $\ln \{\mathbf{A}_3 \exp[\mathbf{A}_2 \ln(\mathbf{A}_1 \mathbf{p}_1)]\}$ by $\mathbf{A}_4$, where

$$\mathbf{A}_4 \atop (2 \times 4) = \begin{bmatrix} 1 & -1 & 0 & 0 \\ 0 & 0 & 1 & -1 \end{bmatrix}$$

We now have $\ln[(A - D)/(A + D)]$. To obtain $(A - D)/(A + D)$, take the exponential of these functions, which produces the desired gammas. The complex function in matrix terms is

$$\mathop{\mathbf{F}}_{(2 \times 1)} = \exp(\mathbf{A}_4 \ln\{\mathbf{A}_3 \exp[\mathbf{A}_2 \ln(\mathbf{A}_1\mathbf{p})]\}) = \begin{bmatrix} 0.88 \\ 0.90 \end{bmatrix}$$

The value $G = 0.88$ computed for the data in panel (*a*) of Table 9.1 is a "partial gamma" in the sense that it is the correlation of response 1 and response 2 *only* for individuals with less than a high school education who twice expressed an opinion about the school integration statement. Since there are three subpopulations defined by education level and two coefficients computed for each subpopulation, there are six estimated gamma coefficients of interest in this problem. These six coefficients, computed by the GENCAT program, are shown in Table 9.4. Because the quantities A, D, T_x, T_y, and T_{xy} can be formed by linear, logarithmic, and exponential operations, analyses using functions such as gamma, tau, or Somers' measures are within the scope of WLS methodology.

Table 9.4   Gamma Coefficients for Data in Table 9.1

Education	Gamma for Expressing Opinion vs. No Opinion	Gamma for Two Agree/ Not Sure/Disagree Responses
Less than high school	0.90	0.88
High school	0.81	0.85
Some college	0.94	0.81

Two problems sometimes arise when one is using matrix manipulation in the GENCAT program to compute complex functions from frequency data. First, the number of functions computed in the intermediate steps to derive the rank correlations is large and can often exceed the storage capacities of many computers. The second problem arises from the fact that it is sometimes necessary to take the natural logarithm of a cell count of zero, but the natural logarithm of zero is undefined. This problem arises in the computation of the gammas from panel (*b*) of Table 9.1. The recommended solution is to substitute a nonzero value—say $1/r$, where r is the number of response categories for the subpopulation. (See Chapter 2 for further discussion of this point.) For this problem $r = 16$ and $1/r = 0.06$. The practical effect on the statistical estimates is negligible: The exact values of the estimated gamma coefficients are 0.85386 and 0.81395, whereas the values computed by GENCAT using the $1/r$ substitution rule are 0.85338 and 0.81401. Analysts may avoid both problems by writing a special computer routine or using software available at a local computer center to calculate $\mathbf{F}$ and $\mathbf{V}_F$, which may then be input directly to the GENCAT program (see Appendix D).

EXAMINING CONSISTENCY

The purpose of computing the six estimated gamma coefficients is to examine the relationship between education and response consistency. The sociological literature suggests that respondents with little education will be least consistent and highly educated respondents will be most consistent (Converse, 1964). Since we are measuring two types of consistency, we also expect a difference between the opinion/no opinion and the agree/disagree coefficients. We begin our investigation with the model

$$\mathbf{X}_1\boldsymbol{\beta}_1 = \begin{bmatrix} 0 & 1 & 0 & -1 \\ 1 & 0 & -1 & 0 \\ 0 & 1 & 0 & 0 \\ 1 & 0 & 0 & 0 \\ 0 & 1 & 0 & 1 \\ 1 & 0 & 1 & 0 \end{bmatrix} \begin{bmatrix} \beta_{10} \\ \beta_{20} \\ \beta_{11} \\ \beta_{21} \end{bmatrix}$$

where β_{10} = mean gamma for agree/disagree response
β_{20} = mean gamma for opinion/no opinion response
β_{11} = coefficient of a linear education term (we have assumed equally spaced levels of education) for agree/disagree response
β_{21} = coefficient of a linear education term for opinion/no opinion response

The linear education term is shown in the $\mathbf{X}$ matrix by -1 if the group has less than a high school education, by 0 if the group has graduated from high school but not attended college, and by 1 if the group has some college education. Note that other values could be assigned here if one does not wish to assume equal spacing for the educational categories.

The test statistic for the fit of this model is $X^2_{\text{GOF}} = 1.55$. This value is less than the seventy-fifth percentile of the chi-square distribution ($\chi^2_{2,0.75} = 2.77$)—a preliminary indication of an adequate fit.

The parameter estimates are

$$\mathbf{b} = \begin{bmatrix} 0.85 \\ 0.91 \\ -0.03 \\ 0.02 \end{bmatrix}$$

The parameter estimates b_{11} and b_{12} suggest that an inverse relationship exists between education and consistency (gamma) for the agree/disagree response ($b_{11} = -0.03$) and a direct relationship between education and consistency for the opinion/no opinion response ($b_{21} = 0.02$). Tests reveal that neither of these effects is significantly different from zero, however. The test statistics for the hypotheses of no linear education effects are $X_1^2 = 1.67$ and $X_1^2 = 0.89$ for the agree/disagree response and the opinion/no opinion response, respectively. These are both less than $\chi_{1,0.95}^2 = 3.84$, so we may comfortably delete these terms from the model. The test statistic for the hypothesis that the mean gammas are not significantly different from one another (H_0: $\beta_{10} = \beta_{20}$) is $X_1^2 = 4.15$. This is greater than the ninety-fifth percentile of the chi-square distribution and hence we reject the hypothesis of no difference.

Because there are no statistically significant education effects, we fit a reduced model that includes β_{10} and β_{20} only. Our original estimates of β_{10} and β_{20} are modified slightly by the presence of the education terms; therefore, to obtain clean estimates not contaminated by the nonsignificant education effects, we fit this reduced model. The model fits well: The test statistic for lack of fit is $X_4^2 = 4.11$, which is less than $\chi_{4,0.75}^2 = 5.39$. The estimates of the mean gammas are $b_{10} = 0.85$ and $b_{20} = 0.92$. These two values differ statistically as the test statistic of the hypothesis of no difference is $X_1^2 = 5.63$ and $\chi_{1,0.95}^2 = 3.84$. Hence the consistency level differs between the agree/disagree response and the opinion/ no opinion response.

RESULTS: NO EDUCATION EFFECT

The data on responses to the school integration question do not support the expectation that people with lower education are less consistent. Both measures of consistency—one measuring the consistency in forming an opinion and the other measuring stability of opinion—show that education had no statistically significant effect on these two measures of association. The estimated gamma coefficient for the formation of an opinion (opinion versus no opinion) is 0.92 and the estimated gamma for opinion stability is 0.85. The data suggest a slight, but not statistically significant, linear effect of education level on the estimated gamma coefficient measuring agree/disagree opinion stability, but the estimated effect is not in the hypothesized direction. Moreover, inspection of the observed gammas for the opinion/no opinion coefficients suggests a possible quadratic relationship between education and consistency. We did not test the result with a strict test, because we had not hypothesized the relationship beforehand.

SUMMARY

This chapter has demonstrated the feasibility of computing measures of association such as the rank correlation coefficient gamma with the WLS methodology. These statistical measures have a wide range of application, especially in the social sciences, but the limitations of space prevent the presentation of further examples.

The key topics covered in this chapter are:

- *The problem:* Determine whether consistency of response to survey questions varies by education level.
- *Rank correlation:* A method of analyzing multiple responses; we used gamma coefficient in this chapter.
- *Consistency:* All the gammas had large values, at least 0.80.
- *Results:* The consistency measures were not affected by education in a linear fashion.

Chapters 8 and 9 have shown three ways of examining multiresponse data. In Chapter 10 we use the mean score approach to a different multiresponse situation: the assignment of ranks.

EXERCISES

9.1.

Analyze the data presented in Exercise 8.1 using the gamma function.

a. Compute the estimated gamma function and the associated variance-covariance matrix for each subtable. Test the hypothesis that there is no linear effect of education on response consistency (as measured by gamma) between the first and second response.

b. Analyze these data using the function "proportion consistent from time 1 to time 2." Test the hypothesis that consistency is not affected by education.

c. Construct a *net change score* for these data that is similar to the function defined in Exercise 6.1. Do you consider this function suitable for this application? Explain your answer. Test the following hypotheses using this function:

1. The net change score is not affected by level of education.
2. The net change score is zero for each level of education.

d. Compare your findings to the analyses performed previously. Which analysis seems best for these data? Explain.

9.2.

The data in Table 9.5 are from a national ulcer study. The response variable is the severity of the *dumping syndrome*, a possible undesirable consequence of surgery for duodenal ulcer. Four hospitals and four surgical procedures are

Table 9.5   Dumping Syndrome by Operation Type and Hospital

HOSPITAL	OPERATION A			OPERATION B			OPERATION C			OPERATION D		
	None	Slight	Mod.	None	Slight	Mod.	None	Slight	Mod.	None	Slight	M
I	23	7	2	23	10	5	20	13	5	24	10	
II	18	6	1	18	6	2	13	13	2	9	15	
III	8	6	3	12	4	4	11	6	2	7	7	
IV	12	9	1	15	3	2	14	8	3	13	6	

Source: Grizzle, Starmer, and Koch (1969).

reported in this problem. The surgical procedures are

   A: drainage and vagotomy

   B: 25 percent resection (antrectomy) and vagotomy

   C: 50 percent resection (hemigastrectomy) and vagotomy

   D: 75 percent resection

Procedure A is least radical and procedure D is most radical.

   a. Form a measure of association between the severity of the dumping syndrome and the surgical procedure for each of the four hospitals.

   b. Does the measure of association depend on the hospital variable? How would you interpret these findings?

   c. What recommendations, if any, would you make to the administrators of the hospitals? What reservations would you place on your recommendations? Why?

   d. Use the mean score approach from Chapter 6 and examine the relationship between a mean severity score and the hospital and type-of-operation variables.

   e. Compare the results of both analyses.

9.3.

In Exercise 7.4 the log cross-product ratio function was used to characterize the association between attitudes toward the candidate and voting preferences at two points in time. Another function characterizing the association is Goodman and Kruskal's gamma (G). Use the estimated gamma function to test whether the association between attitude toward the Republican candidate and voting preference is the same for both time periods.

a. Construct the following functions using GENCAT:

$$G_{vc1} = \text{rank correlation coefficient for voting preference}$$
$$\text{and attitude toward candidate at time 1}$$

$$G_{vc2} = \text{rank correlation coefficient for voting preference}$$
$$\text{and attitude toward candidate at time 2}$$

These functions and their variance-covariance structure may be estimated by using GENCAT, other software, or hand calculation. The following steps illustrate the matrix approach used in the GENCAT program:

1. $A_1 p$ produces F_1.
2. $\ln(F_1)$ produces F_2.
3. $A_2 F_2$ produces F_3.
4. $\exp(F_3)$ produces F_4.
5. $A_3 F_4$ produces F_5.
6. $\ln(F_5)$ produces F_6.
7. $A_4 F_6$ produces F_7.
8. $\exp(F_7)$ produces F_8.

The p vector is the same as in Exercise (7.4).

$$
A_1 =
\begin{bmatrix}
1 & 1 & 1 & 1 & 0 & 0 & 0 & 0 & 0 & 0 & 0 & 0 & 0 & 0 & 0 & 0 \\
0 & 0 & 0 & 0 & 1 & 1 & 1 & 1 & 0 & 0 & 0 & 0 & 0 & 0 & 0 & 0 \\
0 & 0 & 0 & 0 & 0 & 0 & 0 & 0 & 1 & 1 & 1 & 1 & 0 & 0 & 0 & 0 \\
0 & 0 & 0 & 0 & 0 & 0 & 0 & 0 & 0 & 0 & 0 & 0 & 1 & 1 & 1 & 1 \\
1 & 0 & 0 & 0 & 1 & 0 & 0 & 0 & 1 & 0 & 0 & 0 & 1 & 0 & 0 & 0 \\
0 & 1 & 0 & 0 & 0 & 1 & 0 & 0 & 0 & 1 & 0 & 0 & 0 & 1 & 0 & 0 \\
0 & 0 & 1 & 0 & 0 & 0 & 1 & 0 & 0 & 0 & 1 & 0 & 0 & 0 & 1 & 0 \\
0 & 0 & 0 & 1 & 0 & 0 & 0 & 1 & 0 & 0 & 0 & 1 & 0 & 0 & 0 & 1 \\
\end{bmatrix}
\quad
\begin{array}{l}
\textit{row label} \\
a: \text{R+, time 1} \\
b: \text{R−, time 1} \\
c: \text{D+, time 1} \\
d: \text{D−, time 1} \\
a: \text{R+, time 2} \\
b: \text{R−, time 2} \\
c: \text{D+, time 2} \\
d: \text{D−, time 2} \\
\end{array}
$$

A_2 is a 4×8 matrix; A_3 is a 4×4 matrix; A_4 is a 2×4 matrix.

$$
F_8 = \begin{bmatrix} G_{vc1} \\ G_{vc2} \end{bmatrix}
$$

b. Test the following hypotheses:

1. The association between attitude toward the Republican candidate and voting preference at time 1 is zero.

2. The association between attitude toward the Republican candidate and voting preference at time 2 is zero.

3. The association between attitude toward the Republican candidate and voting preference is constant across time periods.

Explain the relationship between tests (1 and 2) and (3) above. If one cannot reject the hypothesis of no association for tests (1) and (2), is it necessary to test (3)?

APPENDIX: GENCAT INPUT

The following listing is for the GENCAT control cards that reproduce the analysis described in this chapter.

```
      5      1      1                          CHAPTER 9: RANK CORRELATION ANALYSIS
      3    16                                  (16F3.0)
  86    4 13    3    5    5    8    3 11    5 74    4    7    2    7 15
  86    5 13    2    5.06    8    2 16    1 61    6    4    4 12 10
 144    5 25    4    7 23 10    1 14    4 67    5    5    4    3 14
      1      2    60    20      1          (20F2.0)
  1 1 1 0 1 1 1 0 1 1 1 0 0 0 0 0
  0 0 0 0 0 0 0 0 0 0 0 0 0 0 0 1
  0 0 0 1 0 0 0 1 0 0 0 1 0 0 0 0
  0 0 0 0 0 0 0 0 0 0 0 1 1 1 0
  1 0 0 0 0 0 0 0 0 0 0 0 0 0 0 0
  0 0 0 0 0 1 1 0 0 1 1 0 0 0 0 0
  0 1 0 0 0 0 0 0 0 0 0 0 0 0 0 0
  0 0 0 0 0 0 1 0 0 0 1 0 0 0 0 0
  0 0 0 0 1 0 0 0 0 0 0 0 0 0 0 0
  0 0 0 0 0 0 0 0 0 1 1 0 0 0 0 0
  0 0 0 0 0 1 0 0 0 0 0 0 0 0 0 0
  0 0 0 0 0 0 0 0 0 1 0 0 0 0 0 0
  0 0 1 0 0 0 0 0 0 0 0 0 0 0 0 0
  0 0 0 0 1 1 0 0 1 1 0 0 0 0 0 0
  0 1 0 0 0 0 0 0 0 0 0 0 0 0 0 0
  0 0 0 1 0 0 0 1 0 0 0 0 0 0 0 0
  0 0 0 0 0 0 1 0 0 0 0 0 0 0 0 0
  0 0 0 0 0 0 0 0 1 1 0 0 0 0 0 0
  0 0 0 0 0 1 0 0 0 0 0 0 0 0 0 0
  0 0 0 0 0 0 0 0 1 0 0 0 0 0 0 0
      2
      1      2    30    10              (20F2.0)
  1 1
 000001  1
 0000000001  1
 00000000000001  1
 000000000000000001  1
 0000000000000000000001  1
 00000000000000000000000001  1
 0000000000000000000000000000001  1
 000000000000000000000000000000000101
 00000000000000000000000000000000000000101
      3
      1      2    12      4              (10F2.0)
 01-1
 0101
```

```
000001010101-1-1-1-1
00000101010101010101
    2
    1      2      6      2              (4F2.0)
01-1
000001-1
    3
    7      1      4                     (6F2.0)
010101010101
01-101-101-1
-10000000100
00-100000001
    8      1      1                     (4F2.0)
0001
    8      1      1                     (4F2.0)
000001
    8      1      1                     (4F2.0)
00000001
    7      1      3                     (6F2.0)
010101010101
01-101-101-1
00-100000001
    8      1      1                     (3F2.0)
000100
    8      1      1                     (3F2.0)
000001
    7      1      2                     (6F2.0)
010101010101
01-101-101-1
    8      1      1                     (3F2.0)
000100
    7      1      1                     (6F2.0)
010101010101
```

10

Rank Choice Analysis

We encounter a new type of multiresponse situation in this chapter. People are asked to *rank* several choices, and the ranks they assign represent the multiple responses. The problem is to test whether there is a clear preference ordering. If the number of choices is large, the corresponding contingency table will have millions of cells. A direct analysis of such a large contingency table is impossible, but the method of analysis discussed in this chapter makes the table unnecessary.

The *rank choice problem* is common to decision making in public health and public affairs. An example of this problem arises during budget allocation. A budgeting authority, such as the U.S. Congress, is faced with a set of spending alternatives—defense, transportation, health, education, and so on—and a fixed resource (dollars) to allocate among them. The questions facing the Congress are not "How much do we spend for defense?" and "How much do we spend for transportation?" but "How much do we spend for defense at the expense of something else?" The last question emphasizes the reality of budgeting—resources are limited but demands on them are not.

The problem central to choice analysis is to determine whether there are differences in preferences for a set of alternatives. In the simplest situation, we may have only two choices—*A* and *B*—and wish to determine whether *A* or *B* is preferred, or whether there is *indifference* between the two alternatives. Rank choice problems differ from other choice situations because of the existence of a constraint—a limit on the possible ordering of alternatives. In our example involving alternatives *A* and *B*, the question is: "Which alternative, *A* or *B*, is preferred?" Posing the question in this manner implies that only one alternative can be preferred and that we cannot have our cake and eat it too.

164

CHOOSING AMONG COMPETING ALTERNATIVES

Rank choice problems differ according to whether the resource can be ranked (ordered) only or whether the resource is continuous and thus apportioned. We refer to the first situation as having a *discrete* resource and the second as having a *continuous* one. Whenever we must decide what to do first or what is "liked" or "preferred," we have a discrete-choice problem. An example of the discrete-choice situation is selecting two flavors of ice cream for a double dip cone. The question is: "Among the 32 flavors, which two do you want?" Continuous-choice situations, however, which are common in management decision making, involve the allocation of a continuous resource—money, time, personnel—to a set of alternatives.

The Houston Community Study introduced in Chapter 9 contains both discrete-choice and continuous-choice situations. The mass opinion questionnaire included a series of questions measuring citizen preference toward programs designed to improve Houston's quality of life. The study's corporate sponsors were interested in setting general priorities among five broad areas of social action and establishing program priorities in each area. The sponsors intended to use the study's findings to develop spending policies for their community service projects.

The five broad policy areas for action were social services, business and job development, public transportation, arts and leisure time activities, and education. For each area, citizens were asked how they would allocate $100,000 among program alternatives. The question that measured preferences about social service program alternatives was:

> Let's suppose you sit on a committee that has been given a large sum of money, say $100,000, to spend on improving social services in Houston. Here is a list of things other people have suggested that the money be spent on. How would you spend $100,000? You should consider spending the money on those suggestions you think are most important. You could spend it all on one thing or divide it up among the various items. Take a minute to look the list over and then tell me how you would spend it.

The nine social service alternatives were:

1. Organize emergency assistance programs.
2. Set up a program to rehabilitate alcoholics.
3. Provide a center for runaway teenagers.
4. Provide a program to help drug addicts.

5. Set up a free family planning clinic.
6. Provide daycare centers for working people.
7. Help the handicapped.
8. Provide care for all elderly people who need it.
9. Other (specify).

After respondents had answered similar questions for each of the other four policy areas, they were asked to *rank* the five areas according to which one was "most important to spend money on." Each of the five program area questions required the allocation of a *continuous* resource ($100,000) to the alternatives for each program area. The data from the social services area are analyzed later in the chapter. In the next section we examine rank choice data for the five areas in order to illustrate the analysis of data using a *discrete* resource.

DISCRETE-CHOICE DATA AND FEW CHOICES

Consider a situation where one must order three alternatives *A*, *B*, and *C*. These alternatives could be as simple as deciding what to do on Saturday night—*A*: watch television at home; *B*: go to a restaurant; or *C*: visit friends—or as complex as choosing among spending alternatives. Given three (*a* in general) alternatives to be ordered, there are 3! (or *a*!) = $3 \cdot 2 \cdot 1 = 6$ possible permutations or preference orderings. Table 10.1 lists these six possible orderings and a hypothetical distribution of 83 people who selected among the alternative orderings. Table 10.1 shows that preference ordering 5 (312)—which means that *A* was ranked third, *B* first, and *C* second—was selected by 36 people in our hypothetical data set. The probabilities $p_{11}, p_{12}, p_{13}, p_{14}, p_{15}$, and p_{16} are the observed proportions of respondents in the sample who selected each ordering. We see from Table 10.1 that $p_{15} = 0.434$.

We may construct a mean rank score for each alternative from these data in a manner similar to that used to construct the mean scores in Chapter 6. The *mean rank score* is a weighted average of the observed probabilities p_{1j} (or just p_j for convenience because we are considering only one population). In general the mean rank for alternative *i* is

$$\bar{r}_i = \sum_{j=1}^{c} w_j p_j$$

where *c* is the number of preference orderings (*c* = 6 in this example, *a*! in general), w_j is the rank that individuals in preference ordering *j* have assigned to alternative *i* (*i* = *A*, *B*, or *C*), and p_j is the observed probability of the preference ordering *j*.

**Table 10.1   Untransformed and Transformed Ranks Assigned
to Hypothetical Data**

Preference Ordering	UNTRANSFORMED RANKS ALTERNATIVE			TRANSFORMED RANKS ALTERNATIVE[a]			Frequency n_{1j}	Probability p_{1j}
	A	B	C	A	B	C		
1	1	2	3	−1	0	1	20	0.241
2	1	3	2	−1	1	0	14	0.169
3	2	1	3	0	−1	1	6	0.072
4	2	3	1	0	1	−1	0	0.000
5	3	1	2	1	−1	0	36	0.434
6	3	2	1	1	0	−1	7	0.084
							83	1.000

Mean preference for alternative A:
$$\bar{r}_A = 1(0.241) + 1(0.169) + 2(0.072) + 2(0.000) + 3(0.434) + 3(0.084) = 2.11$$
Mean preference for alternative B:
$$\bar{r}_B = 2(0.241) + 3(0.169) + 1(0.072) + 3(0.000) + 1(0.434) + 2(0.084) = 1.66$$
Mean preference for alternative C:
$$\bar{r}_C = 3(0.241) + 2(0.169) + 3(0.072) + 1(0.000) + 2(0.434) + 1(0.84) = 2.23$$
Transformed mean preference for alternative A:
$$\bar{r}_A^* = -1(0.241) + (-1)(0.169) + 0(0.072) + 0(0.000) + 1(0.434) + 1(0.084)$$
$$= 0.11$$
Transformed mean preference for alternative B:
$$\bar{r}_B^* = 0(0.241) + 1(0.169) + (-1)(0.072) + 1(0.000) + (-1)(0.434) + 0(0.084)$$
$$= -0.34$$

[a] The transformation is $w_i^* = w_i - 2$, where w_i^* is the transformed rank score for alternative $i (i = 1, 2, 3)$ and w_i is the original rank assigned.

The mean ranks calculated in Table 10.1 are $\bar{r}_A = 2.11$, $\bar{r}_B = 1.66$, and $\bar{r}_C = 2.23$. Note that the sum of the mean ranks equals 6 (the sum of 1, 2, and 3, the three ranks that people can assign). If we know two mean ranks, then we also know the value of the third mean rank. The existence of this relationship causes the variance-covariance matrix of the mean ranks to be singular, which means that we cannot perform further analysis. To avoid this problem we focus our attention on transformed mean ranks and exclude one mean rank score from the analysis. As we will demonstrate, the transformed ranks permit direct testing of the hypotheses of interest, and the exclusion of one rank solves the singularity problem.

The transformation we use subtracts the average rank score of 2 from the original ranks. Thus the new mean ranks are $\bar{r}_A^* = 0.11$, $\bar{r}_B^* = -0.34$, and $\bar{r}_C^* = 0.23$. Note that the transformation has not removed the singularity, however, since these mean ranks sum to zero, and if we know any two mean ranks, we know the third one. To overcome the problem we arbitrarily delete one of the three mean ranks and proceed with the analysis based on only two mean ranks. This elimination of one mean rank does not limit the analysis because knowledge of two ranks ($a - 1$ in general) specifies the value of the third (a-th) rank. The choice of which mean rank to delete is a matter of convenience to the analyst. We include $\bar{r}_A^*$ and $\bar{r}_B^*$ in the following analysis.

The central hypothesis is whether there is no preference among alternatives *A*, *B*, and *C*. This hypothesis, which is sometimes called the *hypothesis of total indifference*, is H_0: $\bar{R}_A = \bar{R}_B = \bar{R}_C$, where $\bar{R}_i$ is the true population mean rank for the *i*th alternative. This hypothesis can be expressed in terms of $\bar{R}_A^*$ and $\bar{R}_B^*$ as H_0: $\bar{R}_A^* = \bar{R}_B^* = 0$; in this case $\bar{R}_i^* = \bar{R}_i - 2$. Note that if both $\bar{R}_A^*$ and $\bar{R}_B^*$ are zero, this implies that $\bar{R}_C^*$ also is zero (because these three mean ranks sum to zero). If all three transformed mean ranks are zero, it means that they are equal to one another. Their equality also implies the equality of the untransformed mean ranks.

DISCRETE-CHOICE DATA AND MANY CHOICES

The previous section illustrated the relationship between frequency data and the functions of interest for testing hypotheses about indifference. We saw that the number of possible orderings depends on the number of alternatives *a*. When *a* alternatives are ranked there are *a*! possible preference orderings, but the number of statistically independent mean rank scores is only $a - 1$. We may illustrate the relationship between the number of preference orderings, the number of alternatives, and the number of mean rank functions as follows:

Number of Alternatives Ranked (a)	Possible Preference Orderings (a!)	Number of Independent Mean Rank Scores (a − 1)
3	6	2
4	24	3
5	120	4
10	3,628,800	9
15	1.308×10^{12}	14

Since the objective of the statistical analysis is to test hypotheses about the mean rank scores, the computation of the vector of these functions **F** and its associated variance-covariance matrix $\mathbf{V}_F$ may proceed directly without resorting to the frequency table from which the functions are derived. This approach is usually necessary because it is impractical to compute the underlying frequency table. The data from the Houston Community Study represent the ranking of five alternatives for which there are 120 possible orderings. It is therefore more efficient to proceed directly with the computation of the vector **F** of the mean rank scores and the variance matrix $\mathbf{V}_F$ using the raw data from the analysis file.

The steps for defining the mean ranks for the five areas are as follows:

1. Assign ranks for the alternatives such that the sum of each respondent's scores is 15 ($= 1 + 2 + 3 + 4 + 5$).

2. Transform the ranks assigned in step 1 by subtracting the *average rank* from each score. The average rank is equal to the sum of the rank scores divided by the number of alternatives *a*. In this case the average rank is 3 (15 divided by 5).

3. Omit one alternative. In the analysis that follows we arbitrarily omit the last alternative (education).

4. Compute the mean rank scores and estimated variance-covariance matrix by using one of two methods:

   a. Use a computer program to estimate **F** and $\mathbf{V}_F$ and input the resulting vector of means and its variance-covariance matrix directly into the GENCAT program.

   b. Use the raw-data input mode in the GENCAT program and define the functions.

We have illustrated both methods of data input in the appendix to this chapter.

Once we have computed the function vector and estimated variance-covariance matrix, the analysis proceeds as described in previous chapters. First we must fit a suitable model, evaluate its fit, and then test hypotheses of indifference under the model. If necessary, we reformulate the model and continue hypothesis testing and interpretation.

The transformed mean rank scores $\bar{r}^*$ and their variance-covariance matrix are given below for the five alternatives from the Houston Community Study data. They are based on an original coding where 1 means first choice and 5 means last choice. Since these original scores were transformed by subtracting 3 from each rank, the actual values of r^* range from -2 to $+2$, where -2 means first choice and $+2$ means last choice.

k	Alternative	$\bar{r}_k$	$\bar{r}_k^*$	$V_F \times 10^{-4}$			
				1	*2*	*3*	*4*
1	Social services	3.01	0.01	15	−3	−5	−2
2	Business/jobs	2.45	−0.55		14	−4	−3
3	Public transportation	3.04	0.04			16	−3
4	Arts/leisure	4.49	1.49				9
5	(Education)	(2.01)	(−0.99)				

We have included the education mean rank score, even though this is the excluded alternative, to illustrate its relationship to the other four mean functions. Observe that the sum of the original ranks is 15 and that

$$\bar{r}_5 = 15 - \sum_{k=1}^{4} \bar{r}_k$$

Similarly,

$$\bar{r}_5^* = - \sum_{k=1}^{4} \bar{r}_k^*$$

The mean rank scores reveal that education is the preferred area of spending ($\bar{r}_5^* = -0.99$), followed by business and job development (-0.55), social services (0.01), public transportation (0.04), and arts and leisure activities (1.49).

We now wish to determine whether there is a significant difference in the mean ranks—that is, the hypothesis of total indifference. This hypothesis is expressed as:

$$H_0: \quad \bar{R}_1^* = 0$$
$$\bar{R}_2^* = 0$$
$$\bar{R}_3^* = 0$$
$$\bar{R}_4^* = 0$$

where $\bar{R}_i^*$ is the true population transformed mean rank estimated by $\bar{r}_i^*$. This hypothesis is generated by considering the saturated model

$$\mathbf{F}_{(4 \times 1)} = \begin{bmatrix} \bar{r}_1^* \\ \bar{r}_2^* \\ \bar{r}_3^* \\ \bar{r}_4^* \end{bmatrix} = \begin{bmatrix} 1 & 0 & 0 & 0 \\ 0 & 1 & 0 & 0 \\ 0 & 0 & 1 & 0 \\ 0 & 0 & 0 & 1 \end{bmatrix} \begin{bmatrix} \beta_1 \\ \beta_2 \\ \beta_3 \\ \beta_4 \end{bmatrix}$$

This saturated model fits perfectly, which allows us to proceed with the test of hypothesis of total indifference. The $\mathbf{C}$ matrix that generates this hypothesis is $\mathbf{C} = \mathbf{I}_4$. The test statistic for this hypothesis is $X^2 = 3236.26$, which is greater than the critical value of the chi-square distribution; therefore, we reject the hypothesis of total indifference.

If this test were not statistically significant, we could end the analysis at this point because there would be little variation in the mean ranks. Since this is not the case, we may examine additional hypotheses to determine which ranks differ significantly from one another. Comparing the mean rank for social services with the other alternatives yields the following results:

ypothesis	Description	Contrast Matrix $\mathbf{C}$				X^2	DF
$\bar{R}_1^* = \bar{R}_2^*$	Social services − business/jobs	[1	−1	0	0]	86.71	1
$\bar{R}_1^* = \bar{R}_3^*$	Social services − public transportation	[1	0	−1	0]	.26	1
$\bar{R}_1^* = \bar{R}_4^*$	Social services − arts/leisure	[1	0	0	−1]	773.55	1

These tests suggest that there is no difference in preference for social services and public transportation (alternatives 1 and 3).

Using this information, we fit a reduced model $\mathbf{X}_2\boldsymbol{\beta}_2$, where

$$\mathbf{X}_2\boldsymbol{\beta}_2 = \begin{bmatrix} 1 & 0 & 0 \\ 0 & 1 & 0 \\ 1 & 0 & 0 \\ 0 & 0 & 1 \end{bmatrix} \begin{bmatrix} \beta_1 \\ \beta_2 \\ \beta_3 \end{bmatrix}$$

The β_1 parameter is the transformed mean rank score for social services and public transportation; β_2 is the mean rank for business and job development; β_3 is for arts and leisure activities. The goodness-of-fit X^2_{GOF} is 0.26 (DF = 1). This value is the same as the test statistic for the hypothesis $\bar{R}_1^* = \bar{R}_3^*$—which we expect, given the meaning of the X^2_{GOF} with DF = 1. (See Chapter 4 for another example.)

We may test hypotheses under this model as before. The test results suggest that the social services/public transportation transformed mean rank (0.02) is significantly different from the business and job development mean rank (−0.55) and the arts and leisure score (1.49). One additional hypothesis of interest concerns the education mean, the alternative omitted. We may test whether $\bar{R}_5^*$, the transformed mean rank for education, differs significantly from the business/job mean—that is, H_0: $\bar{R}_5^* - \bar{R}_2^* = 0$—by using the contrast matrix $\mathbf{C} = \begin{bmatrix} -2 & -2 & -1 \end{bmatrix}$. This matrix is the difference between the matrices $\begin{bmatrix} -2 & -1 & -1 \end{bmatrix}$ and $\begin{bmatrix} 0 & 1 & 0 \end{bmatrix}$, which represent the $\mathbf{C}$ matrices testing the hypotheses respectively that $\bar{R}_5^*$ and $\bar{R}_2^*$ equal zero. The first parameter estimate for the hypothesis H_0: $\bar{R}_5^* = 0$ is counted twice in the matrix $\begin{bmatrix} -2 & -1 & -1 \end{bmatrix}$, because the education mean rank score is the negative sum of the estimates for the other four alternatives and the social service and transportation alternatives are both estimated by the first term. The test statistics for the hypothesis H_0: $\bar{R}_5^* - \bar{R}_2^* = 0$ is $X^2 = 58.75$. Thus we see that a statistically significant difference exists between the mean rank scores for education and business and job development.

We may summarize the estimated preference orderings based on model $\mathbf{X}_2\boldsymbol{\beta}_2$ as follows:

Alternative	Estimated Mean Rank	Estimated Transformed Mean Rank	Order
Education	2.01	−0.99	First choice
Business/job development	2.45	−0.55	Second choice
Social services } Public transportation }	3.02	0.02	Third choice (tied)
Arts/leisure activities	4.49	1.49	Last choice

We found partial indifference on spending between the social services and public transportation, but otherwise there is a clear preference for spending in other areas. Education was preferred over all other choices, and spending for arts and leisure time was least preferred.

CONTINUOUS-CHOICE DATA

In continuous-choice problems, the resource to be allocated may be subdivided into relatively fine intervals. The allocation of $100,000 among nine social service programs is but one example of a continuous-choice problem.

The Houston Community Study required the respondent to allocate $100,000 in units of $1000 or more across nine alternatives. The contingency table representing the possible number of allocation choices is very large indeed. As a consequence, most rank choice analyses, regardless of whether the resource is discrete or continuous, involve contingency tables having more cells (orderings) possible than sample observations. Although this fact may appear startling at first, it does not compromise rank choice analysis because the functions analyzed are weighted sums computed across the entire table. As a consequence, they are based on the *total* sample for the subpopulation.

The continuous-choice problem, regardless of the alternatives presented, allows the analyst to choose between a traditional parametric approach such as *multivariate analysis of variance* (MANOVA) and the WLS method. Many social analysts would treat these data as nine interdependent mean scores sampled from a single population. To analyze these data using a MANOVA approach requires us to assume that the scale (dollars allocated) is continuous and that the population is distributed approximately as multivariate normal. The WLS method does not assume that the data are continuous, but it does require the analyst to justify the weighting scheme used to construct mean scores. Nor does the WLS method require assumptions about the underlying distribution. Because of the asymptotic nature of the test statistics and because the mean scores are based on a relatively large sample (over 900 cases), we can comfortably assume that the X^2 test statistics approximately follow the chi-square distribution. The choice between methods is partly philosophical, in the sense that the analyst may prefer one position or the other, but the problem's substance may dictate which method is better suited for analyzing the data. We now illustrate the categorical data analysis.

We compute the vector of eight transformed mean rank scores and its estimated variance-covariance matrix directly from the raw data. The data are first scored such that each respondent's total allocation sums to $100 ($\times 1000$). The scores are transformed by subtracting $11.11 (100/9 = 11.11) to center them around zero. The last alternative ("other") is excluded arbitrarily.

Table 10.2 gives the observed transformed mean scores $\bar{r}_k^*$ and their estimated variance-covariance matrix. These data suggest clear preferences among the population sampled in allocating money for the elderly care (12.43 = $12,430) and handicapped (7.91 = $7,910) programs over the remaining alternatives. The total-indifference test, based on model $X_3 \beta_3$ ($X_3 = I_8$), shows $X^2 = 544.83$ (DF = 8)—an indication that statistically significant differences do in fact exist among the nine mean spending scores.

We therefore begin an investigation to determine whether differences between the individual means exist and to discover whether there is a simplified preference ordering. Based on Table 10.2 five alternatives—emergency assistance (1), alcoholic rehabilitation (2), runaway teen program (3), family planning clinic (5), and daycare center (6)—appear to have similar levels of support and appear to be different from drug addiction program (4). These six alternatives may be ordered from least support to most support by using their observed

Table 10.2 Observed Transformed Mean Scores $\bar{r}^*$ and Estimated Variance-Covariance Matrix V_F for Social Services Data

Alternative	Observed $\bar{r}_k^*$	VARIANCE-COVARIANCE MATRIX V_F ($\times 10^{-4}$)							
		1	2	3	4	5	6	7	8
1. Emergency assistance	−2.73	2302	−78	−97	−271	−152	−345	−370	−728
2. Alcoholic rehabilitation	−2.48		2530	−50	−276	−355	−416	−618	−1018
3. Runaway teen program	−3.80			2027	−109	−228	−198	−575	−552
4. Drug addiction program	−0.01				3122	−418	−370	−769	−1180
5. Family planning clinic	−2.14					3094	−99	−637	−876
6. Daycare center	−1.81						3096	−647	−725
7. Handicapped program	7.91							4761	−557
8. Elderly program	12.43								6391
9. (Other)[a]	(−7.37)								

[a] The $\bar{r}_9^*$ ("other") mean score equals the negative sum of the first eight means.

mean rank scores as follows:

k			Program	$\bar{r}_k^*$
Least support				
Test e	$\begin{cases} 3 \\ 1 \\ 2 \\ 5 \end{cases}$	Test h	Runaway teen program	-3.80
			Emergency assistance	-2.73
			Alcoholic rehabilitation	-2.48
			Family planning clinic	-2.14
Test f	$\begin{cases} 6 \\ 4 \end{cases}$		Daycare center	-1.81
			Drug addiction program	-0.01
Most support				

Tests a to h, reported in Table 10.3, show the relationships of the spending scores for the five alternatives (1), (2), (3), (5), and (6). The preceding summary highlights the alternatives compared by tests e, f, and h.

Tests a to d in Table 10.3 demonstrate that alternatives 2, 3, 5, and 6 are not statistically different from alternative 1, the alternative approximately halfway between alternatives 3 and 6. These individual tests suggest that there is no preference ordering among these five alternatives. Test e shows that simultaneous tests of the first four alternatives (1, 2, 3, and 5) do not differ. The chi-square statistic is 6.37, which is not significant at the $\alpha = 0.01$ level. The difference between the next two alternatives (4 and 6) is tested by contrast f. This test shows that these two alternatives differ at the $\alpha = 0.05$ but not the $\alpha = 0.01$ levels.

Since test d has already suggested that alternatives 1 and 6 are not statistically different and test g shows that alternative 6 is not statistically different from a weighted average of alternatives 1, 2, 3, and 5, we now test whether alternative 6 may be grouped with the four alternatives tested by test e. The results of the test that alternatives 1, 2, 3, 5, and 6 are statistically similar is shown in test h of Table 10.3. This simultaneous test of the first five alternatives is not statistically significant at the $\alpha = 0.05$ level ($\chi^2_{5,\,0.95} = 9.49$). This test means that the alternatives in the range from -3.80 for the runaway teen program to -1.81 for the daycare center program receive the same level of support. We would thus say that indifference exists in the preference ordering for these five alternatives. Finally tests i and j show that the remaining alternatives 4, 7, and 8 are different from each other.

Using the information in Table 10.3, we choose a reduced model containing a single parameter to estimate alternatives 1, 2, 3, 5, and 6. We know from test h of Table 10.3 that the goodness-of-fit X^2 will equal 9.09. Had we estimated alternative 6 with a separate parameter, the goodness-of-fit X^2 would have

Table 10.3 Summary of Selected Tests of Significance for Mean Spending Scores

Hypothesis	C Matrix	X^2	DF	Estimate
a. $\bar{R}_1^* - \bar{R}_2^* = 0$	$[1\ \ -1\ \ 0\ \ 0\ \ 0\ \ 0\ \ 0]$	0.13	1	−0.26
b. $\bar{R}_1^* - \bar{R}_3^* = 0$	$[1\ \ 0\ \ -1\ \ 0\ \ 0\ \ 0\ \ 0]$	2.49	1	1.06
c. $\bar{R}_1^* - \bar{R}_5^* = 0$	$[1\ \ 0\ \ 0\ \ 0\ \ -1\ \ 0\ \ 0]$	0.62	1	−0.60
d. $\bar{R}_1^* - \bar{R}_6^* = 0$	$[1\ \ 0\ \ 0\ \ 0\ \ 0\ \ -1\ \ 0]$	1.40	1	−0.92
e. First 3 tests (a–c) equal zero	$\begin{bmatrix} 1 & -1 & 0 & 0 & 0 & 0 & 0 \\ 1 & 0 & -1 & 0 & 0 & 0 & 0 \\ 1 & 0 & 0 & 0 & -1 & 0 & 0 \end{bmatrix}$	6.37	3	−0.26 1.06 −0.60
f. $\bar{R}_4^* - \bar{R}_6^* = 0$	$[0\ \ 0\ \ 0\ \ 1\ \ 0\ \ -1\ \ 0]$	4.64	1	1.80
g. $\bar{R}_{1235}^* - \bar{R}_6^* = 0$	$[0.25\ \ 0.25\ \ 0.25\ \ 0\ \ 0.25\ \ -1\ \ 0]$	2.30	1	−0.97
h. Tests (a–d) all equal zero	$\begin{bmatrix} 1 & -1 & 0 & 0 & 0 & 0 & 0 \\ 1 & 0 & -1 & 0 & 0 & 0 & 0 \\ 1 & 0 & 0 & 0 & -1 & 0 & 0 \\ 1 & 0 & 0 & 0 & 0 & -1 & 0 \end{bmatrix}$	9.09	4	−0.26 1.06 −0.60 −0.92
i. $\bar{R}_4^* - \bar{R}_7^* = 0$	$[0\ \ 0\ \ 0\ \ 1\ \ 0\ \ 0\ \ -1]$	66.63	1	−7.92
j. $\bar{R}_7^* - \bar{R}_8^* = 0$	$[0\ \ 0\ \ 0\ \ 0\ \ 0\ \ 0\ \ 1\ \ -1]$	16.64	1	−4.52

been 6.37 (see test *e*). We choose the more parsimonious model and leave open the question of whether an independent study will confirm the correctness of our choice.

The reduced model $\mathbf{X}_4\boldsymbol{\beta}_4$ is

$$
\begin{bmatrix}
1 & 0 & 0 & 0 \\
1 & 0 & 0 & 0 \\
1 & 0 & 0 & 0 \\
0 & 1 & 0 & 0 \\
1 & 0 & 0 & 0 \\
1 & 0 & 0 & 0 \\
0 & 0 & 1 & 0 \\
0 & 0 & 0 & 1
\end{bmatrix}
\begin{bmatrix}
\beta_1 \\
\beta_2 \\
\beta_3 \\
\beta_4
\end{bmatrix}
$$

The parameter β_1 represents the common amount allocated to alternatives 1, 2, 3, 5, and 6. Parameters β_2, β_3, and β_4 represent the amount allocated to alternatives 4, 7, and 8, respectively. The following contrasts demonstrate that all the parameters are statistically different from one another:

Hypothesis	Contrast Matrix $\mathbf{C}$	X^2	DF	Estimated Difference
$\beta_1 - \beta_2 = 0$	$\begin{bmatrix} 1 & -1 & 0 & 0 \end{bmatrix}$	19.69	1	-2.70
$\beta_2 - \beta_3 = 0$	$\begin{bmatrix} 0 & 1 & -1 & 0 \end{bmatrix}$	67.10	1	-7.89
$\beta_3 - \beta_4 = 0$	$\begin{bmatrix} 0 & 0 & 1 & -1 \end{bmatrix}$	17.57	1	-4.63

We may also test whether the excluded alternative (other programs) is equal to β_1 (the five pooled alternatives) using the $\mathbf{C}$ matrix: $\mathbf{C} = \begin{bmatrix} -6 & -1 & -1 & -1 \end{bmatrix}$. This matrix is based on $\bar{r}_9^* = -\sum_{i=1}^{8} \bar{r}_i^*$. Since β_1 represents the value of five of the alternatives, it has a coefficient of -5 and the other three β's have a coefficient of -1. To test the hypothesis H_0: $\bar{R}_9^* - \beta_1 = 0$, we must also subtract another β_1; hence the -5 becomes -6. The test statistic for this hypothesis is $X^2 = 56.00$ and the estimate of $\bar{R}_9^* - \beta_1 = -4.67$. The test shows that the "other programs" alternative is allocated an average of \$4670 (-4.67) less than the amount allocated the runaway teen program (3), emergency assistance (1), alcoholic rehabilitation (2), family planning (5), and a daycare center (6).

We have summarized the estimates of mean spending preference based on model $\mathbf{X}_4\boldsymbol{\beta}_4$ in Table 10.4 and transformed these estimates into dollars.

**Table 10.4   Estimates of Spending Preferences for Social Services Programs
Based on Reduced Model $X_4\beta_4$**

Alternative	*Estimated β_i Based on Model $X_4\beta_4$*	*Transformed Estimate*[a]
Care for elderly	12.57	$23,680
Help handicapped	7.94	19,050
Help drug addicts	0.05	11,160
Organize emergency assistance		
Rehabilitate alcoholics		
Provide for runaway teens	−2.65	8,460
Set up family planning clinics		
Provide daycare centers		
Other programs	−7.32	3,790

[a] Transformed estimate = $(b_i + 11.11) \times \$1000$.

Table 10.4 shows that the nine alternatives are only partially ordered with three programs—care for the elderly, aid to the handicapped, and drug assistance—clearly preferred over the remaining six spending alternatives. We also note that even though respondents were given the opportunity to propose additional spending alternatives (the "other" alternative), they were not likely to allocate much money to them.

The conclusions from this analysis must be interpreted with caution because our analysis was guided by the data. Since we did not have a theory for a guide, the tests we performed were selected according to the values of the parameters. One way of dealing with this problem is to split the data randomly into two groups. Then analyze one group and use the other to verify the results of the first analysis. This procedure presumes that the data set is large. Since we did not believe that the data set analyzed here had a large enough sample to allow us to split the data into two groups, our findings are limited to this data set. They may actually apply to a more general population, but this point requires additional verification before we can safely conclude that.

SUMMARY

This chapter illustrated the generality of rank choice analysis. The situations range from the complete ranking of a small set of alternatives to the allocation of continuous resources across many alternatives. In each case, the data may be conceived as having an underlying contingency table, based on one sub-population, in which each cell represents a distinct ordering of the alternatives.

At one extreme, the table can be small, which means that the number of cells is less than the number of subjects. At the other extreme, the number of cells substantially exceeds the sample size. Regardless of the size of the table, one is usually interested in estimating a small number of functions based on these cells and not in analyzing the probability of a single preference ordering. The function most often used is a mean score.

Rank choice analysis is not confined to the analysis of data from a single population. The approach may be generalized to include two or more sub-populations, but the addition of data from more than one subpopulation, though adding to the complexity of the analysis, does not change its essential character.

The key topics covered in this chapter are:

- *The problem:* Rank five program areas for funding; allocate funds to nine alternatives within the social service program.
- *Discrete-choice data (few choices):* Assign ranks to the categories in the table and form mean scores.
- *Discrete-choice data (many choices):* The contingency table is not required; form mean scores and the estimated variance-covariance matrix directly; use direct-input option.
- *Continuous-choice data (many choices):* Alternative to MANOVA; weights must be selected; contingency table need not be formed.

This chapter completes our discussion of multiple responses. In the next chapter we present the WLS approach to follow-up life tables—a methodology created for drug comparisons but applicable to situations in which it is necessary to track people over time. This application provides another example of a complex function.

EXERCISES

10.1.

Table 10.5, based on the Houston Community Study data, presents an expansion of the problem presented in the text by introducing information on the preferences of whites and blacks regarding social service programs. Determine whether the preferences of whites and blacks are the same and whether there is indifference among a subset of alternatives by considering the following:

a. Examine the mean rank scores $\bar{r}$ for each subpopulation. Do the orderings appear similar? Do there appear to be any significant differences between whites and blacks regarding the relative importance of some programs? Explain your answer.

Table 10.5 Observed Transformed Mean Scores $\bar{r}^*$ and Estimated Variance-Covariance Matrices V_F for Social Services by Race

Alternative	Observed $\bar{r}^*$	VARIANCE-COVARIANCE MATRIX V_F $(\times 10^{-4})$							
		1	2	3	4	5	6	7	8
Whites									
1. Emergency assistance	-2.51	3150	-131	-194	-354	-163	-462	-339	-1038
2. Alcoholic rehabilitation	-2.45		3745	-168	347	-330	-591	-880	-1517
3. Runaway teen program	-3.49			3059	-105	-194	-378	-882	-731
4. Drug addiction program	-0.04		(symmetric)		4356	-589	-498	-1057	-1588
5. Family planning clinic	-2.58					3965	-209	-870	-1055
6. Daycare center	-2.11						4759	-929	-1165
7. Handicapped program	7.46							6757	-769
8. Elderly program	12.15								9257
Blacks									
1. Emergency assistance	-3.49	13549	-109	534	-84	344	-2373	-5225	-5675
2. Alcoholic rehabilitation	-1.42		13993	1126	1439	-2347	-2814	-4478	-5730
3. Runaway teen program	-5.36			6118	-64	-611	4	-2693	-3283
4. Drug addiction program	-0.02		(symmetric)		16413	-1848	-2633	-5016	-6676
5. Family planning clinic	-4.56					10393	383	-187	-5302
6. Daycare center	-1.05						13156	-3076	-1812
7. Handicapped program	9.74							24995	-2278
8. Elderly program	15.18								32638

Source: Houston Community Study data; Stafford and Lehnen (1975).

b. Input the function vector **F** and estimated variance-covariance matrix V_F for blacks only to the GENCAT program using the case 3 option. Fit the final model $X_4\beta_4$ presented in this chapter for social service data and evaluate its fit. Compare your findings with those reported in this chapter for the full sample.

c. Input the entire 16-DF problem into the GENCAT program. Test the 8-DF hypothesis of no difference in preference orderings for whites and blacks. Do you think that other models should be fit to these data? Explain your conclusions.

d. Try an alternative model that includes the main effects for question ordering and race (9-DF model). Evaluate its fit and interpret the results. What have you learned about the preferences of Houstonians for social service programs?

10.2.

Koch, Freeman, and Lehnen (1976) studied the attitudes of a national sample of Americans toward cutting taxes versus spending for government programs. Respondents were asked to rank seven tax policy alternatives (descending order from 1 to 7)—spending for education (ED), control of water and air pollution (PL), antipoverty programs (PV), foreign aid (FA), guaranteed minimum income (GI), health care (HC), or returning tax money in the form of a tax reduction (TR). Respondents were classified according to their ideology (conservative, liberal, in-between, no ideology), sex (male, female), and criticism of government tax policies (no, yes). Table 10.6 gives data on the observed mean rank preference function for the seven tax policy alternatives for 4 of the 16 subpopulations analyzed by Koch, Freeman, and Lehnen.

a. Analyze the data from the male, conservative, no criticism subpopulation only and test the hypothesis of indifference. Is there a clear preference among the alternatives? *Hint:* One alternative must be excluded from the analysis and the data must be transformed by subtraction of a constant. What is the constant?

b. Perform an analysis similar to part (*a*) but compare the preference orderings of male and female conservatives who expressed no criticism of tax policies. Does the sex of the respondent affect the ordering of tax policy alternatives for conservatives who did not criticize the government?

c. Compare the four subpopulations for which data are given and answer the following questions:

1. Is there a similar tax policy preference among the four subpopulations?

2. What role do the respondent's sex and ideology play in shaping attitudes toward government tax policy?

Table 10.6 Preference Data for Tax Policy Alternatives by Sex and Ideology

	GOVERNMENT PROGRAMS						
	ED	PL	TR	PV	FA	GI	HC

Males with conservative ideology
who expressed no tax criticism ($n_1 = 155$)

	ED	PL	TR	PV	FA	GI	HC
Mean rank preference vector:	2.1816	3.5966	3.0109	5.1105	6.1992	4.6499	3.2512
Variance-covariance matrix:	0.0157	0.0001	−0.0058	0.0007	−0.0007	−0.0083	−0.0017
	0.0001	0.0178	−0.0013	−0.0037	−0.0011	−0.0068	−0.0048
	−0.0058	−0.0013	0.0214	−0.0063	−0.0001	−0.0034	−0.0042
	0.0007	−0.0037	−0.0063	0.0134	−0.0010	−0.0016	−0.0013
	−0.0007	−0.0011	−0.0001	−0.0010	0.0082	−0.0032	−0.0018
	−0.0083	−0.0068	−0.0034	−0.0016	−0.0032	0.0199	0.0036
	−0.0017	−0.0048	−0.0042	−0.0013	−0.0018	0.0036	0.0105

Females with conservative ideology
who expressed no tax criticism ($n_2 = 121$)

	ED	PL	TR	PV	FA	GI	HC
Mean rank preference vector:	2.0944	3.4947	3.0528	4.9129	6.2066	4.6758	3.5628
Variance-covariance matrix:	0.0152	−0.0012	−0.0059	0.0015	−0.0002	−0.0088	−0.0005
	−0.0012	0.0247	0.0005	−0.0038	−0.0001	−0.0131	−0.0069
	−0.0059	0.0005	0.0311	−0.0093	−0.0040	−0.0061	−0.0061
	0.0015	−0.0038	−0.0093	0.0159	−0.0011	−0.0013	−0.0019
	−0.0002	−0.0001	−0.0040	−0.0011	0.0114	−0.0044	−0.0013
	−0.0088	−0.0131	−0.0061	−0.0013	−0.0044	0.0294	0.0044
	−0.0005	−0.0069	−0.0061	−0.0019	−0.0013	0.0044	0.0125

Males with liberal ideology
who expressed no tax criticism ($n_3 = 84$)

Mean rank preference vector:

2.3855	4.1743	3.7608	4.1796	6.2644	4.1105	3.1249

Variance-covariance matrix:

0.0331	-0.0028	-0.0041	-0.0040	0.0011	-0.0185	-0.0046
-0.0028	0.0279	-0.0021	-0.0073	-0.9016	-0.0095	-0.0044
-0.0041	-0.0021	0.0571	-0.0207	-0.0159	-0.0089	-0.0052
-0.0040	-0.0073	-0.0207	0.0389	0.0091	-0.0045	-0.0113
0.0011	-0.0016	-0.0159	0.0091	0.0130	-0.0026	-0.0029
-0.0185	-0.0095	-0.0089	-0.0045	-0.0026	0.0392	0.0048
-0.0046	-0.0044	-0.0052	-0.0113	-0.0029	0.0048	0.0238

Females with liberal ideology
who expressed no tax criticism ($n_4 = 76$)

Mean rank preference vector:

2.2112	4.0107	4.7372	3.8395	6.1061	3.6574	3.4379

Variance-covariance matrix:

0.0272	-0.0067	-0.0061	0.0024	0.0048	-0.0090	-0.0126
-0.0067	0.0433	0.0027	-0.0083	-0.0024	-0.0241	-0.0045
-0.0061	0.0027	0.0574	-0.0212	-0.0042	-0.0192	-0.0093
0.0024	-0.0083	-0.0212	0.0334	0.0004	-0.0018	-0.0049
0.0048	-0.0024	-0.0042	0.0004	0.0134	-0.0058	-0.0062
-0.0090	-0.0241	-0.0192	-0.0018	-0.0058	0.0528	0.0073
-0.0126	-0.0045	-0.0093	-0.0049	-0.0062	0.0073	0.0304

3. How much can you generalize from these data? What limitations, if any, should you place on interpretation of the findings?

Note: The numerical results derived from the direct input of **F** and $\mathbf{V}_F$ reported in this exercise may vary slightly from the values reported in Koch, Freeman, and Lehnen (1976) because of errors introduced by rounding **F** and $\mathbf{V}_F$.

10.3.

The German National Social Survey, a survey of social attitudes conducted in 1980 and replicated annually in West Germany by the Center for Surveys, Methods, and Analysis (ZUMA), produced data on German attitudes toward "materialistic" versus "postmaterialistic" values (Table 10.7). Respondents were presented four statements and asked to order them from most important (rank 1) to least important (rank 4) for their country in the next 10 years. The equivalent English statements are:

A: Preserve law and order in this land.

B: Provide a greater role of the citizen in governmental decisions.

C: Fight against rising prices.

D: Protect the right of free speech.

Statements B and D are defined as "postmaterialist" values; statements A and C represent "materialist" values.

a. Determine whether there is no difference in the ordering of the four statements within each of the nine education and age subpopulations. Construct a suitable **A** matrix that defines the mean rank score for statements A, B, and C. This matrix will be of size (27 × 216) and have a block-diagonal structure:

$$\mathbf{A}_{(27 \times 216)} = \begin{bmatrix} \mathbf{A}^* & & & & & & & & \\ & \mathbf{A}^* & & & & & & & \\ & & \mathbf{A}^* & & & & & & \\ & & & \mathbf{A}^* & & & & & \\ & & & & \mathbf{A}^* & & & & \\ & & & & & \mathbf{A}^* & & & \\ & & & & & & \mathbf{A}^* & & \\ & & & & & & & \mathbf{A}^* & \\ & & & & & & & & \mathbf{A}^* \end{bmatrix}$$

where $\mathbf{A}^*$ is a (3 × 24) submatrix and the off-diagonal values are zero. *Hint:* The sum of the ranks 1, 2, 3, and 4 is 10; thus the "raw" ranks must be centered by subtracting 2.5 from each score.

Table 10.7 Ordering of Four Questions Pertaining to Postmaterialism and Materialism Values by Age and Education

STATEMENT RANK[a]	LOW			MEDIUM			HIGH		
Education:[b]									
ABCD — Age:	<35	35–60	60+	<35	35–60	60+	<35	35–60	60+
1234	33	62	54	13	27	17	3	6	5
1243	22	35	15	14	12	6	4	5	3
1324	46	98	112	12	20	26	2	4	5
1342	13	31	14	16	22	10	5	6	3
1423	58	130	140	13	32	28	1	8	5
1432	25	56	26	14	20	14	4	3	5
2134	12	26	19	8	12	7	4	3	0
2143	11	14	8	10	7	3	4	5	3
2314	32	62	49	5	17	11	3	6	1
2341	8	18	2	11	9	5	8	3	1
2413	30	59	41	16	11	4	4	1	0
2431	8	19	5	9	12	4	5	1	1
3124	12	20	10	3	4	3	3	0	0
3142	19	13	4	14	17	0	11	7	0
3214	22	26	22	8	6	1	2	2	0
3241	13	15	7	17	9	2	17	3	0
3412	12	31	14	10	9	2	1	0	2
3421	9	13	3	10	9	1	1	1	1
4123	11	10	6	13	7	3	8	4	1
4132	15	11	5	19	12	1	26	9	1
4213	13	20	4	5	5	4	1	0	1
4231	17	14	9	20	13	2	34	10	2
4312	10	18	7	3	6	1	5	2	0
4321	8	8	3	12	5	4	4	3	0

Source: German National Social Survey, 1980. The authors wish to thank Professor Dr. Manfred Kuechler, Frankfurt University, West Germany, for suggesting this exercise and providing the frequency table.

[a] "Statement rank" refers to the ranking of statements A, B, C, and D. Thus the value "2413" means that A ranked second, B was fourth, C was first, and D was third. There are 4! = 24 ways to order four statements.

Fit a saturated model having the following structure:

$$
\mathbf{X}_{(27 \times 27)} =
\begin{bmatrix}
\mathbf{X}^* & & & \\
& \mathbf{X}^* & & \\
& & \cdots & \\
& & & \mathbf{X}^*
\end{bmatrix}
$$

where $\mathbf{X}^* = \mathbf{I}_3$. Each submatrix $\mathbf{X}^*$ accounts for the variation in question ordering within an education $\times$ age subpopulation. If a set of three parameter estimates is not statistically significant, then one can conclude that indifference exists regarding the choice of statements for that subpopulation. Test whether there is indifference regarding the statements within each of the nine subpopulations.

b. The principal substantive hypothesis regarding the acceptance of post-materialistic over materialistic values may be expressed as H_0: $\bar{R}_A + \bar{R}_C = \bar{R}_B + \bar{R}_D$, or in other words, the sum of the mean rank scores for statements A and C equals the sum of the mean rank scores for statements B and D. This hypothesis may be characterized by a function of the data in Table 10.7 of the form

$$
f_i = \bar{r}_A - \bar{r}_B + \bar{r}_C - \bar{r}_D
$$

where i represents the subpopulation ($i = 1,9$). A test that the function equals zero is a test of indifference between the sets of materialistic and post-materialistic statements.

Construct this function for each of the nine subpopulations and test whether there is indifference in the choice among the two types of statements. *Hint:* The required $\mathbf{A}$ matrix will be of size (9×216).

APPENDIX: GENCAT INPUT

Discrete-Choice Data

The following listing is for the GENCAT control cards that reproduce the discrete-choice analysis described in this chapter. The analysis is based on the "raw" data.

```
     5    4    1              INPUT OF DISCRETE DATA FOR CHAP 10
     1    1    4    4         (4F4.2,4X1,F4.2)
S(1)=G(1)
F(1)=MEAN(1)
F(2)=MEAN(2)
F(3)=MEAN(3)
F(4)=MEAN(4)
ORDER=(D,D,D,D,I)
```

```
   0-200-100 100 200 100
   0-200 100 200-100 100
-100 100-200 200    0 100
 200 100-200    0-100 100
 200 100    0-100-200 100
```

[Data Cards Omitted]

```
-100    0 100 200-200 100
 100    0-200 200-100 100
-100 100    0 200-200 100
 200    0 100-100-200 100
 100    0-100 200-200 100
0000000000000000000-999
        7    4    4           1                    IDENTITY DESIGN MATRIX
        8    1    1                  (4F2.0)         1 - 2
01-1
        8    1    1                  (4F2.0)         1 - 3
0100-1
        8    1    1                  (4F2.0)         1 - 4
010000-1
        6                    REANALYSIS TO FIT REDUCED MODEL
        7    1    3                  (4F2.0)           REDUCED MODEL
01000100
0001
00000001
        8    1    1                  (3F2.0)         1 - 2
01-1
        8    1    1                  (3F2.0)         2 - 3
0001-1
        8    1    1                  (3F2.0)         5 - 2
-2-2-1
```

Alternative Input

The following is an alternative listing of GENCAT control cards using the direct input of **F** and $\mathbf{V}_F$ estimated by other computer programs. The statistics derived from this analysis may vary slightly from the values reported in the chapter because of the rounding of **F** and $\mathbf{V}_F$.

```
        5    3    1           CHAP 10: ANALYSIS OF DISCRETE CHOICE DATA
        4                            (4F10.4)
.0078       -.5458     .0401  1.4932
        1
.0012          -.0003        -.0005         -.0002
-.0003          .0014        -.0004         -.0003
-.0005         -.0004         .0016         -.0003
-.0002         -.0003        -.0003          .0009
        7    4    4           1                    IDENTITY DESIGN MATRIX
        8    1    1                  (4F2.0)         1 - 2
01-1
        8    1    1                  (4F2.0)         1 - 3
0100-1
        8    1    1                  (4F2.0)         1 - 4
010000-1
        6                    REANALYSIS TO FIT REDUCED MODEL
        7    1    3                  (4F2.0)           REDUCED MODEL
```

```
01000100
0001
00000001
     8       1      1                    (3F2.0)              1  -  2
01-1
     8       1      1                    (3F2.0)              2  -  3
0001-1
     8       1      1                    (3F2.0)              5  -  2
-2-2-1
```

Continuous-Choice Data

The following listing is for the GENCAT control cards that reproduce the
continuous-choice analysis described in this chapter. The analysis is based on
the "raw" data.

```
     5      4      1                     INPUT OF CONTINUOUS DATA FOR CHAP 10
     1      1      8      8              (8F6.2,6X1,F6.2)
S(1)=G(1)
F(1)=MEAN(1)
F(2)=MEAN(2)
F(3)=MEAN(3)
F(4)=MEAN(4)
F(5)=MEAN(5)
F(6)=MEAN(6)
F(7)=MEAN(7)
F(8)=MEAN(8)
ORDER=(D,D,D,D,D,D,D,D,I)
       89     189      89     189      89     189      89     189  -1111     100
       89     189      89     189      89     189      89     189  -1111     100
       89     189      89     189      89     189      89     189  -1111     100
    -1111   -1111   -1111   -1111   -1111   -1111   -1111   -1111   8889     100
     1389   -1111   -1111   -1111    1389    1389    1389   -1111  -1111     100
```

[Data Cards Omitted]

```
    -1111   -1111   -1111   -1111   -1111   -1111    3889    3889  -1111     100
    -1111   -1111    8889   -1111   -1111   -1111   -1111   -1111  -1111     100
    -1111    1389   -1111   -1111    1389   -1111    1389    1389  -1111     100
    -1111   -1111    -111   -1111    3889   -1111     889     889  -1111     100
    -1111    8889   -1111   -1111   -1111   -1111   -1111   -1111  -1111     100
                                                                           -9999
       7      4      8                    1                    IDENTITY MATRIX
       8      1      1                    (8F2.0)              1  -  2
01-1
       8      1      1                    (8F2.0)              1  -  3
0100-1
       8      1      1                    (8F2.0)              1  -  5
01000000-1
       8      1      1                    (8F2.0)              1  -  6
0100000000-1
       8      1      4                    (8F2.0)              1  -  2,3,5,6  TEST
01-1
0100-1
01000000-1
0100000000-1
       8      1      3                    (8F2.0)              1  -  2,5,6  TEST   DF=3
01-1
01000000-1
0100000000-1
```

```
      8      1      1              (8F2.0)            1 - 3
0100-1
      8      1      1              (8F2.0)            4 - 6
0000000100-1
      8      1      1              (8F2.0)            4 - 7
000000010000-100
      8      1      1              (8F2.0)            7 - 8
00000000000001-1
      8      1      1              (8F3.2)            R1,2,5,6 - R3
0250251.0000025025
      8      1      1              (8F3.2)            R1,2,3,5 - R6
025025025000025-1.
      6                      1     REANALYSIS TO FIT REDUCED MODEL
      7      1      4              (8F2.0)            REDUCED MODEL
010101000101
00000001
00000000000001
0000000000000001
      8      1      1              (4F2.0)            1 - 2
01-1
      8      1      1              (4F2.0)            2 - 3
0001-1
      8      1      1              (4F2.0)            3 - 4
000001-1
      8      1      1              (4F2.0)            R9=0
-5-1-1-1
      8      1      1              (4F2.0)            R12356=0
01
      8      1      1              (4F2.0)            R9 - R12356
-6-1-1-1
      6                            REANALYSIS TO FIT REDUCED MODEL
      7      1      5              (8F2.0)            REDUCED MODEL TWO
0101010001
00000001
000000000001
00000000000001
0000000000000001
      8      1      1              (5F2.0)            R1,2,3,5=R6
0100-1
```

Alternative Input

The following is an alternative listing of GENCAT control cards using the direct input of **F** and **V**$_F$ estimated by other computer programs. The statistics derived from this analysis may vary slightly from the values reported in this chapter because of the rounding of **F** and **V**$_F$.

```
      5      3      1              CHAP 10: ANALYSIS OF CONTINUOUS CHOICE DATA
      8      1      8              (8F10.4)'
-2.7258    -2.5090    -3.5978    -.0774    -2.0985    -1.7411    7.87      12.2388
      2                            (8F10.4)
.2194      -.0072     -.0104     -.0257    -.0133     -.0326     -.0353    -.0698
.2401      -.0049      .027      -.0335    -.0391     -.0596     -.0969
.2027      -.0108     -.0232     -.0197    -.0555     -.0566
.2966      -.0399     -.0345     -.0736    -.1118
.2956      -.0088     -.0616     -.0847
.2959      -.0628     -.0695
.4549      -.0502
.6105
```

```
     7    4
     8    1    1              (8F2.0)           1 - 2
01-1
     8    1    1              (8F2.0)           1 - 3
 0100-1
     8    1    1              (8F2.0)           1 - 5
01000000-1
     8    1    1              (8F2.0)           1 - 6
0100000000-1
     8    1    4              (8F2.0)           1 - 2,3,5,6 TEST
01-1
0100-1
01000000-1
0100000000-1
     8    1    3              (8F2.0)           1 - 2,5,6 TEST   DF=3
01-1
01000000-1
0100000000-1
     8    1    1              (8F2.0)           1 - 3
0100-1
     8    1    1              (8F2.0)           4 - 6
0000000100-1
     8    1    1              (8F2.0)           4 - 7
000000010000-100
     8    1    1              (8F2.0)           7 - 8
00000000000001-1
     8    1    1              (8F3.2)           R1,2,5,6 - R3
0250251.0000025025
     8    1    1              (8F3.2)           R1,2,3,5 - R6
025025025000025-1.
     6                    1  REANALYSIS TO FIT REDUCED MODEL
     7    1    4              (8F2.0)           REDUCED MODEL
010101000101
00000001
00000000000001
0000000000000001
     8    1    1              (4F2.0)           1 - 2
01-1
     8    1    1              (4F2.0)           2 - 3
0001-1
     8    1    1              (4F2.0)           3 - 4
000001-1
     8    1    1              (4F2.0)           R9=0
-5-1-1-1
     8    1    1              (4F2.0)           R12356=0
01
     8    1    1              (4F2.0)           R9 - R12356
-6-1-1-1
     6                        REANALYSIS TO FIT REDUCED MODEL
     7    1    5              (8F2.0)           REDUCED MODEL TWO
0101010001
00000001
000000000001
00000000000001
0000000000000001
     8    1    1              (5F2.0)           R1,2,3,5-R6
0100-1
```

11

Follow-Up Life Table Analysis

In this chapter we demonstrate how the analysis of people's experience over time—for example, survival after some treatment intervention—can be performed with the WLS approach. The analysis of survival data using *life table methodology* has been widely applied in the medical field, especially in clinical trial experiments. Examples of this type of analysis are provided by Merrell and Shulman (1955) and by Zubrod and others (1960). In these studies patients receiving different treatments are followed over time and their survival experience or remission periods are compared. This same methodology is now being used in other policy fields. Potter (1966) studied the retention of contraceptives using life table analysis, for example, while Wollstadt, Shapiro, and Bice (1978) and Forthofer, Glasser, and Light (1980) applied this method to study membership retention in health maintenance organizations (HMOs). The life table approach is applicable whenever one tracks people over time—for example, studying the recidivism of parolees or juvenile delinquents.

In this chapter we examine the membership retention experience in an HMO to demonstrate the application of life table methodology in the WLS framework. Our discussion follows closely that of Koch, Johnson, and Tolley (1972), though Gehan and Siddiqui (1973) provide an alternative format for the analysis of survival data.

PREDICTING HMO MEMBERSHIP

We have seen in Chapter 6 that the use of health services is related to characteristics of the family unit; therefore, for planning purposes, an HMO requires estimates of both the composition and the retention of its membership. The life table approach provides a direct way of examining the issue of membership retention and, in conjunction with WLS methodology, it allows one to assess the impact of particular variables on the time-related retention probabilities.

The data used in this chapter come from the Kaiser Health Services Research Center (Oregon Region), the same source that supplied the data in Chapter 6. In this chapter we focus on membership records and do not include information about use of the Kaiser Plan's services. The data represent the experiences of subscribers in the 5 percent random sample who joined the plan during 1967–1970. We are using the subscriber's experience to represent that of the entire subscriber unit. The subscriber's retention experience, measured by the presence or absence of membership, is tracked and reported on by 6-month intervals.

We could examine a number of factors or variables and their relationship to membership retention, but the small sample size (2115 subscribers) prohibits the simultaneous inclusion of a large number of variables in the analysis. For this analysis we have chosen two variables: age at enrollment and year of enrollment in the plan. The age variable was selected to serve as a proxy for job stability because the younger population is more apt to change jobs whereas the middle-aged and older populations are more likely to remain on their job and hence in the plan. Research by Pope (1978) suggests the importance of job stability in predicting membership retention. The year of enrollment variable reflects possible changes in the plan's service delivery—for example, a declining physician/membership ratio—that may affect a member's satisfaction and thus the decision to retain membership.

The age variable has three levels: less than 25, 25 to 44, and greater than or equal to 45. We would prefer additional levels for age, but the small sample size prohibits further categorization. The year of enrollment variable has four levels: 1967, 1968, 1969, and 1970.

THE FOLLOW-UP LIFE TABLE APPROACH

The *follow-up life table* tracks people's experience over time. In creating this table we must decide on an exact reference point from which to start measuring time and collect the exact follow-up data on the member's experience with the plan, here the decision to retain membership.

Figure 11.1 shows the type of data required. In Figure 11.1*a* we have displayed the experience of five Kaiser Plan members in chronological fashion, whereas in Figure 11.1*b* the same members are presented against a common reference point. The reference point from which we measure experience is the moment an individual enrolls in the plan.

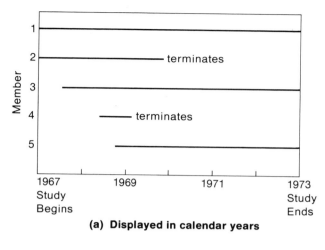

(a) **Displayed in calendar years**

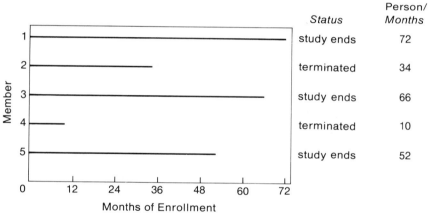

(b) **Adjusted to common starting point**

Figure 11.1 **Hypothetical Membership of Five Members Enrolling in Two Successive Years Followed Over a 6-Year Study Period**

There are three possible membership states at any time x. These states and their mathematical symbols are defined as follows:

O_x = number of people who are still plan members at beginning of time period x.

D_x = number of people who have formally terminated membership during period x.

W_x = number of people who are still plan members at beginning of period x when study terminated. This state also includes cases for whom the administrative system cannot provide their enrollment status (missing data).

To summarize this information, we calculate the following statistics:

- q_x: the estimated probability of termination during time period x:

$$q_x = \frac{D_x}{O_x - 0.5W_x}$$

- p_x: the estimated probability of continuing as a plan member into the next time period:

$$p_x = 1 - q_x$$

- P_x: the estimated probability of a member remaining in the plan from the reference point to time period x:

$$P_x = \prod_{i=1}^{x-1} p_i$$

The calculation of q_x is complicated by the W_x because we do not know whether they would have remained in the plan or dropped out had we been able to observe them after the end of the study. The usual procedure in this case is to compromise by weighting the W_x by 0.5 in the denominator. This weighting of W_x by 0.5 is a recognition that the people in the W_x state were not eligible for termination for the entire period. Since we do not know exactly how long they were eligible, we arbitrarily assume they were eligible for one-half of the period. A survival rate P_x is usually the statistic of central interest; for example, the 5-year survival rate is used in a cancer study by Koch, Johnson, and Tolley (1972).

WLS AND THE FOLLOW-UP LIFE TABLE

Koch, Johnson, and Tolley (1972) showed how the WLS approach could be applied to follow-up life tables for one or more population groups. In the usual presentation of follow-up life tables, the values of O_x, W_x, D_x, q_x, p_x, and P_x are shown for all the values of x. The essential elements, however, are W_x, D_x, and the number alive at the end of the last period. The other statistics can be calculated from these. Table 11.1 contains the essential data for the calculation of the survival or retention rate in this chapter.

Table 11.1 Retention Experience of 2115 Subscribers Who Joined During 1967–1970 by 6-Month Intervals

Group Enrolled During	Age	D_x							W_x							O_{42}	Total
		0	6	12	18	24	30	36	0	6	12	18	24	30	36		
1967	<25	24	16	12	12	5	8	5	0	0	0	0	0	0	0	71	153
	25–44	23	27	20	11	5	12	5	0	0	0	0	0	0	0	111	214
	≥45	8	17	7	9	6	11	2	0	0	0	0	0	0	0	72	132
1968	<25	41	28	14	8	3	2	5	0	0	0	0	0	0	0	74	175
	25–44	45	33	10	12	10	8	7	0	0	0	0	0	0	0	102	227
	≥45	18	13	10	5	3	5	3	0	0	0	0	0	0	0	67	124
1969	<25	56	20	17	7	8	13	3	0	0	0	0	0	0	15	44	183
	25–44	61	33	14	13	11	12	8	0	0	0	0	0	0	43	78	273
	≥45	20	13	9	5	4	4	6	0	0	0	0	0	0	18	48	127
1970	<25	43	24	11	9	8	1	0	0	0	0	0	29	20	7	0	152
	25–44	51	34	10	15	13	0	0	0	0	0	0	38	53	30	0	244
	≥45	16	8	7	8	0	0	0	0	0	0	0	25	24	23	0	111

The period of observation for this study ran from January 1967 through February 1973. We report results only for the subscriber's first 36 months in the plan since the variable of interest here is the 3-year retention rate. The entries under the W_x category are a consequence of being unable to observe people who joined in late 1969 or during 1970 for 3 years.

Note that Table 11.1 is identical in structure to other contingency tables analyzed in previous chapters. There are 12 populations (formed by the combination of age and year) and 15 possible response categories (seven time periods of 6 months in length for both W_x and D_x plus the number of subscribers still on the plan after 42 months). A subscriber appears in one and only one of the 15 categories; thus the categories are mutually exclusive and exhaustive.

To estimate the 3-year retention rate for comparison among the 12 age by year of enrollment groups, we must form P_{36}, which requires that we calculate the p_i for $i = 0, 6, \ldots, 30$ because

$$P_{36} = \prod_{i=0}^{30} p_i$$

Recall that $p_x = 1 - q_x$ and $q_x = D_x/(O_x - 0.5W_x)$. Another way of expressing p_x is

$$\frac{O_x - 0.5W_x - D_x}{O_x - 0.5W_x}$$

To find P_{36} we first form p_x in stages:

1. Form a linear function of the probabilities corresponding to the cells in Table 11.1 by premultiplying **p** (the vector of observed probabilities) by an **A** matrix. The rows of **A** will be sets of two rows with the first row of each set forming the probability of the numerator of p_x $(O_x - 0.5W_x - D_x)$ and the second row forming the probability of the denominator of p_x $(O_x - 0.5W_x)$. Since there are six 6-month time periods (0, 6, 12, 18, 24, 30) to be estimated, **A*** will have 12 rows.

2. After performing the linear transformation in step 1, take the logarithms of the numerators and denominators and form a linear function of these logarithms:

   ln(probability of numerator) − ln(probability of denominator)

   which is equal to ln(probability of numerator/probability of denominator).

3. Finally, take the exponential of this function, which gives the ratio

$$\frac{\text{Probability of numerator}}{\text{Probability of denominator}} = p_x$$

To demonstrate these operations, we first find

$$p_0 = \frac{O_0 - 0.5W_0 - D_0}{O_0 - 0.5W_0}$$

for the first population (1967 enrollment and <25 years old). Recall that if we sum the D_x, W_x, and O_{42} in the first row we get O_0. We now consider the first two rows of $\mathbf{A}^*$:

$$\begin{bmatrix} 0 & 1 & 1 & 1 & 1 & 1 & 1 & 0.5 & 1 & 1 & 1 & 1 & 1 & 1 & 1 \\ 1 & 1 & 1 & 1 & 1 & 1 & 1 & 0.5 & 1 & 1 & 1 & 1 & 1 & 1 & 1 \end{bmatrix}$$

The first row of $\mathbf{A}^*$, when postmultiplied by $\mathbf{p}$, sums the probabilities of all the D_x, the W_x, and O_{42} except for D_0 and $0.5W_0$, which is simply the probability of $(O_0 - D_0 - 0.5W_0)$. The second row of $\mathbf{A}^*$, when postmultiplied by $\mathbf{p}$, gives the probability of $(O_0 - 0.5W_0)$. The same idea applies to finding the numerator and denominator of the p_x's for the other periods.

The numerator and denominator for p_6 are

$$\frac{O_6 - 0.5W_6 - D_6}{O_6 - 0.5W_6}$$

where O_6 represents the people who enter the second 6-month period. The quantity O_6 is equal to the people who are in the study, O_0, minus the people who left during the first 6 months, D_0, and minus those who were enrolled less than 6 months when the study terminated, W_0—that is, $O_6 = O_0 - D_0 - W_0$. The probability of remaining for period $x = 6$ is thus

$$p_6 = \frac{O_0 - D_0 - W_0 - 0.5W_6 - D_6}{O_0 - D_0 - W_0 - 0.5W_6} \tag{11.1}$$

The two rows of $\mathbf{A}_1$, which form the numerator and denominator for p_6, are

$$\begin{bmatrix} 0 & 0 & 1 & 1 & 1 & 1 & 1 & 0 & 0.5 & 1 & 1 & 1 & 1 & 1 & 1 \\ 0 & 1 & 1 & 1 & 1 & 1 & 1 & 0 & 0.5 & 1 & 1 & 1 & 1 & 1 & 1 \end{bmatrix}$$

The first row when postmultiplied by **p** gives the probability of $(O_0 - D_0 - D_6 - W_0 - 0.5W_6)$, which equals the probability of $(O_6 - D_6 - 0.5W_6)$. The second row creates the probability of $(O_0 - D_0 - W_0 - 0.5W_6)$, which equals the probability of $(O_6 - 0.5W_6)$. Thus we have the elements necessary to find p_6. The entire **A*** matrix for the first population is

$$
\mathbf{A^*} =
\begin{bmatrix}
0 & 1 & 1 & 1 & 1 & 1 & 1 & 0.5 & 1 & 1 & 1 & 1 & 1 & 1 & 1 \\
1 & 1 & 1 & 1 & 1 & 1 & 1 & 0.5 & 1 & 1 & 1 & 1 & 1 & 1 & 1 \\
0 & 0 & 1 & 1 & 1 & 1 & 1 & 0 & 0.5 & 1 & 1 & 1 & 1 & 1 & 1 \\
0 & 1 & 1 & 1 & 1 & 1 & 1 & 0 & 0.5 & 1 & 1 & 1 & 1 & 1 & 1 \\
0 & 0 & 0 & 1 & 1 & 1 & 1 & 0 & 0 & 0.5 & 1 & 1 & 1 & 1 & 1 \\
0 & 0 & 1 & 1 & 1 & 1 & 1 & 0 & 0 & 0.5 & 1 & 1 & 1 & 1 & 1 \\
0 & 0 & 0 & 0 & 1 & 1 & 1 & 0 & 0 & 0 & 0.5 & 1 & 1 & 1 & 1 \\
0 & 0 & 0 & 1 & 1 & 1 & 1 & 0 & 0 & 0 & 0.5 & 1 & 1 & 1 & 1 \\
0 & 0 & 0 & 0 & 0 & 1 & 1 & 0 & 0 & 0 & 0 & 0.5 & 1 & 1 & 1 \\
0 & 0 & 0 & 0 & 1 & 1 & 1 & 0 & 0 & 0 & 0 & 0.5 & 1 & 1 & 1 \\
0 & 0 & 0 & 0 & 0 & 0 & 1 & 0 & 0 & 0 & 0 & 0 & 0.5 & 1 & 1 \\
0 & 0 & 0 & 0 & 0 & 1 & 1 & 0 & 0 & 0 & 0 & 0 & 0.5 & 1 & 1 \\
\end{bmatrix}
\begin{array}{l}
\left.\begin{array}{l} \\ \\ \end{array}\right\} p_0 \\
\left.\begin{array}{l} \\ \\ \end{array}\right\} p_6 \\
\left.\begin{array}{l} \\ \\ \end{array}\right\} p_{12} \\
\left.\begin{array}{l} \\ \\ \end{array}\right\} p_{18} \\
\left.\begin{array}{l} \\ \\ \end{array}\right\} p_{24} \\
\left.\begin{array}{l} \\ \\ \end{array}\right\} p_{30}
\end{array}
$$

To construct the **A** matrix for 12 subpopulations, we repeat the matrix for each of the 12 population subgroups as follows:

$$
\underset{(144 \times 180)}{\mathbf{A}} =
\begin{bmatrix}
\mathbf{A^*} & & & & & & & & & & & \\
& \mathbf{A^*} & & & & & & & & & & \\
& & \mathbf{A^*} & & & & & & & & & \\
& & & \mathbf{A^*} & & & & & & & & \\
& & & & \mathbf{A^*} & & & & & & & \\
& & & & & \mathbf{A^*} & & & & & & \\
& & & & & & \mathbf{A^*} & & & & & \\
& & & & & & & \mathbf{A^*} & & & & \\
& & & & & & & & \mathbf{A^*} & & & \\
& & & & & & & & & \mathbf{A^*} & & \\
& & & & & & & & & & \mathbf{A^*} & \\
& & & & & & & & & & & \mathbf{A^*}
\end{bmatrix}
$$

which is the familiar block-diagonal structure.

Since we wish to form the ratio of the two rows created from **A**, we take the natural logarithm of **Ap** and then premultiply the natural logarithms by

$$
\begin{bmatrix} 1 & -1 & 1 & -1 & 1 & -1 & 1 & -1 & 1 & -1 & 1 & -1 \end{bmatrix}
$$

for each population:

$$[1 \quad -1 \quad 1 \quad -1 \quad 1 \quad -1 \quad 1 \quad -1 \quad 1 \quad -1 \quad 1 \quad -1]
\begin{bmatrix}
\ln\{\mathrm{prob}(O_0 - 0.5W_0 - D_0)\} \\
\ln\{\mathrm{prob}(O_0 - 0.5W_0)\} \\
\ln\{\mathrm{prob}(O_6 - 0.5W_6 - D_6)\} \\
\ln\{\mathrm{prob}(O_6 - 0.5W_6)\} \\
\ln\{\mathrm{prob}(O_{12} - 0.5W_{12} - D_{12})\} \\
\ln\{\mathrm{prob}(O_{12} - 0.5W_{12})\} \\
\ln\{\mathrm{prob}(O_{18} - 0.5W_{18} - D_{18})\} \\
\ln\{\mathrm{prob}(O_{18} - 0.5W_{18})\} \\
\ln\{\mathrm{prob}(O_{24} - 0.5W_{24} - D_{24})\} \\
\ln\{\mathrm{prob}(O_{24} - 0.5W_{24})\} \\
\ln\{\mathrm{prob}(O_{30} - 0.5W_{30} - D_{30})\} \\
\ln\{\mathrm{prob}(O_{30} - 0.5W_{30})\}
\end{bmatrix}$$

As a result of this stage we have now formed $\ln(p_0) + \ln(p_6) + \ln(p_{12}) + \ln(p_{18}) + \ln(p_{24}) + \ln(p_{30})$ since

$$\ln\{\mathrm{prob}(O_x - 0.5W_x - D_x)\} - \ln\{\mathrm{prob}(O_x - 0.5W_x)\}$$
$$= \ln\left(\frac{\mathrm{prob}(O_x - 0.5W_x - D_x)}{\mathrm{prob}(O_x - 0.5W_x)}\right) = \ln(p_x)$$

and

$$\ln(p_0) + \ln(p_6) + \ln(p_{12}) + \ln(p_{18}) + \ln(p_{24}) + \ln(p_{30})$$
$$= \ln(p_0 \times p_6 \times p_{12} \times p_{18} \times p_{24} \times p_{30})$$
$$= \ln(P_{36})$$

The exponential of $\ln(P_{36})$ produces P_{36}, the probability of remaining in the plan for 36 months, for each of the 12 year $\times$ age enrollment subpopulations. These probabilities may be analyzed in exactly the same manner as in previous chapters to assess the effects of age and year of enrollment on the probability of retention.

SPECIFYING THE MODEL

We now perform the various steps required to estimate P_{36} for each of the 12 populations. The steps are:

1. Premultiply the 15-element vector **p** by the **A*** matrix shown above.
2. Take the natural logarithm of **A*p**.

3. Premultiply $\ln(\mathbf{A}^*\mathbf{p})$ by the 1×12 matrix

$$[1 \quad -1 \quad 1 \quad -1 \quad 1 \quad -1 \quad 1 \quad -1 \quad 1 \quad -1 \quad 1 \quad -1]$$

4. Take the exponential of the result from step 3.

We perform these steps for each of the 12 populations.

The observed retention rates are shown in Table 11.2 under "3-Year Retention". These rates tell us that approximately half the subscribers will drop the plan within 3 years of their date of enrollment. This figure varies by age: the youngest group has the lowest probability of retention (approximately 34 percent in 1969 and 1970) compared to the oldest group's retention probability of about 60 percent. No great differences in the average retention rate by year of enrollment exist, although 1967 appears to have a slightly higher rate.

To examine the effects of age and year of enrollment more thoroughly, we examine a model that includes the main effects of both these variables. If we define the vector of parameters $\boldsymbol{\beta}_1$ as

$$\boldsymbol{\beta}_1 = \begin{bmatrix} \text{constant} \\ \text{differential effect of 1967} \\ \text{differential effect of 1968} \\ \text{differential effect of 1969} \\ \text{differential effect of age } <25 \\ \text{differential effect of age } 25\text{--}44 \end{bmatrix}$$

the corresponding design matrix $\mathbf{X}_1$ is

$$\mathbf{X}_1 = \begin{bmatrix} 1 & 1 & 0 & 0 & 1 & 0 \\ 1 & 1 & 0 & 0 & 0 & 1 \\ 1 & 1 & 0 & 0 & -1 & -1 \\ 1 & 0 & 1 & 0 & 1 & 0 \\ 1 & 0 & 1 & 0 & 0 & 1 \\ 1 & 0 & 1 & 0 & -1 & -1 \\ 1 & 0 & 0 & 1 & 1 & 0 \\ 1 & 0 & 0 & 1 & 0 & 1 \\ 1 & 0 & 0 & 1 & -1 & -1 \\ 1 & -1 & -1 & -1 & 1 & 0 \\ 1 & -1 & -1 & -1 & 0 & 1 \\ 1 & -1 & -1 & -1 & -1 & -1 \end{bmatrix}$$

Table 11.2 Summary of Model Fitting for the 12 Populations

POPULATION		Number of Subscribers	3-Year Retention	Estimated SE	Predicted Based on $X_1\beta_1$	Predicted Based on $X_2\beta_2$	Predicted Based on $X_3\beta_3$	Estimated SE for $X_3\beta_3$	Residuals Based on $X_3\beta_3$
Year	Age								
1967	<25	153	0.50	0.040	0.45	0.46	0.45	0.023	0.04
	25–44	214	0.54	0.034	0.54	0.49	0.49	0.011	0.05
	≥45	132	0.56	0.043	0.62	0.52	0.54	0.024	0.02
1968	<25	175	0.45	0.038	0.41	0.44	0.45	0.023	−0.00
	25–44	227	0.48	0.033	0.50	0.49	0.49	0.011	−0.01
	≥45	124	0.56	0.045	0.59	0.55	0.54	0.024	0.03
1969	<25	183	0.34	0.035	0.37	0.37	0.36	0.022	−0.02
	25–44	273	0.47	0.030	0.46	0.49	0.49	0.011	−0.02
	≥45	127	0.57	0.044	0.55	0.62	0.63	0.024	−0.06
1970	<25	152	0.34	0.043	0.40	0.34	0.36	0.022	−0.02
	25–44	244	0.49	0.033	0.49	0.49	0.49	0.011	−0.01
	≥45	111	0.65	0.045	0.58	0.65	0.63	0.024	0.02
Fit of the model:					$X_6^2 = 11.19$	$X_7^2 = 6.46$	$X_9^2 = 7.38$		

The statistic for testing the fit of this model is $X^2_{GOF} = 11.19$. Comparing this to the seventy-fifth percentile of the chi-square distribution with 6 degrees of freedom, $\chi^2_{6, 0.75} = 7.84$, we reject the goodness of fit for this model. This result provides some evidence of interaction between age and year of enrollment. Instead of adding additional interaction parameters, we will try another approach to modeling based on our initial thoughts and observations about the data.

We observe that the average 3-year retention rate for each year appears to be similar, but the effect of age, although directly related to the rate, appears to be different within year of enrollment. A modular model reflecting these observations is $X_2 \beta_2$, where β_2 is defined as

$$\beta_2 = \begin{bmatrix} \text{constant} \\ \text{linear age effect for 1967} \\ \text{linear age effect for 1968} \\ \text{linear age effect for 1969} \\ \text{linear age effect for 1970} \end{bmatrix} = \begin{bmatrix} \beta_0 \\ \beta_1 \\ \beta_2 \\ \beta_3 \\ \beta_4 \end{bmatrix}$$

and

$$X_2 = \begin{bmatrix} 1 & -1 & 0 & 0 & 0 \\ 1 & 0 & 0 & 0 & 0 \\ 1 & 1 & 0 & 0 & 0 \\ 1 & 0 & -1 & 0 & 0 \\ 1 & 0 & 0 & 0 & 0 \\ 1 & 0 & 1 & 0 & 0 \\ 1 & 0 & 0 & -1 & 0 \\ 1 & 0 & 0 & 0 & 0 \\ 1 & 0 & 0 & 1 & 0 \\ 1 & 0 & 0 & 0 & -1 \\ 1 & 0 & 0 & 0 & 0 \\ 1 & 0 & 0 & 0 & 1 \end{bmatrix}$$

In this model the differences in age rates within enrollment year are characterized by a simple linear term using the assignment of -1 to the less than 25 age group, 0 to the 25–44 age group, and 1 to the 45 and over group. This assignment is done separately for each of the 4 years of enrollment to permit estimation of different age effects for each year of enrollment.

The goodness-of-fit statistic for $X_2 \beta_2$ is $X_{GOF}^2 = 6.46$, which is less than $\chi_{7,0.75}^2 = 9.04$, and we fail to reject this model. The estimate of β_2 is

$$b_2 = \begin{bmatrix} 0.494 \\ 0.030 \\ 0.054 \\ 0.123 \\ 0.154 \end{bmatrix}$$

The estimated average 3-year retention rate is 0.494. The linear trend estimates appear to be different among the enrollment years; the largest age effects occur during 1969 and 1970. The test of the hypothesis of no difference in the linear trends can be written in a number of ways. One formulation of this hypothesis is

$$H_0: \quad \beta_1 = \beta_2 \qquad \text{or} \qquad H_0: \quad \beta_1 - \beta_2 = 0$$
$$\beta_1 = \beta_3 \qquad\qquad\qquad \beta_1 - \beta_3 = 0$$
$$\beta_1 = \beta_4 \qquad\qquad\qquad \beta_1 - \beta_4 = 0$$

The C matrix that creates this hypothesis is

$$C = \begin{bmatrix} 0 & 1 & -1 & 0 & 0 \\ 0 & 1 & 0 & -1 & 0 \\ 0 & 1 & 0 & 0 & -1 \end{bmatrix}$$

Because of the after-the-fact nature of the analysis, we perform this test at the $\alpha = 0.01$ level. The test statistic for this hypothesis is $X_3^2 = 11.52$, which is greater than the ninety-ninth percentile of the chi-square distribution with 3 degrees of freedom, $\chi_{3,0.99}^2 = 11.34$. Hence we reject the hypothesis of no difference in the linear trends.

It appears that the linear trends for 1967 and 1968 are similar to one another and different from the other two trends. Since the trends for 1969 and 1970 also appear to be similar, we test the hypotheses of no differences between the trends in 1967 and 1968 and between those in 1969 and 1970. These are the hypotheses:

$$H_0: \quad \beta_1 = \beta_2$$

and

$$H_0: \quad \beta_3 = \beta_4$$

The test statistics for these two hypotheses are $X_1^2 = 0.36$ and $X_1^2 = 0.56$, respectively, both of which are less than $\chi_{1,0.99}^2 = 6.63$; therefore we fail to reject the hypotheses of no difference.

We consequently fit a model that has the same linear trend for 1967 and 1968 and the same trend for 1969 and 1970. The parameter vector $\boldsymbol{\beta}_3$ is

$$\boldsymbol{\beta}_3 = \begin{bmatrix} \text{constant} \\ \text{linear trend 1967--1968} \\ \text{linear trend 1969--1970} \end{bmatrix}$$

and

$$\mathbf{X}_3 = \begin{bmatrix} 1 & -1 & 0 \\ 1 & 0 & 0 \\ 1 & 1 & 0 \\ 1 & -1 & 0 \\ 1 & 0 & 0 \\ 1 & 1 & 0 \\ 1 & 0 & -1 \\ 1 & 0 & 0 \\ 1 & 0 & 1 \\ 1 & 0 & -1 \\ 1 & 0 & 0 \\ 1 & 0 & 1 \end{bmatrix}$$

This model fits well as $X_{\text{GOF}}^2 = 7.38$, and this test statistic is less than the critical value $\chi_{9,0.75}^2 = 11.39$. The parameter estimates are

$$\mathbf{b}_3 = \begin{bmatrix} 0.495 \\ 0.042 \\ 0.137 \end{bmatrix}$$

The linear trend in 1967--1968 appears to be only one-third as large as that in 1969--1970. The test of whether this difference is statistically significant is

$$H_0: \quad \beta_1 = \beta_2$$

and the test statistic is $X_1^2 = 10.60$, which is greater than $\chi_{1,0.99}^2 = 6.63$. Hence the two trends are not equal at the $\alpha = 0.01$ level.

Table 11.2 summarizes the results from this analysis. The model fits well, although it underestimates the retention rate for 1967 for all three age groups as shown in the last column of the table. We can also see that the estimated standard errors (SE) for the predicted retention rates are smaller than those of the original estimates in the "Estimated SE" column. These estimated standard errors are smaller because we have used all 2115 observations in arriving at our predicted results; the estimated standard errors for the original estimates are based on each age by year combination considered separately and these sample sizes range from 111 to 273, values much smaller than 2115.

AGE AND YEAR OF ENROLLMENT

Analysis of the 3-year retention rates shows that age is an important variable. It appears that age is operating mostly in a linear fashion, but the coefficient of the linear age trend varies by year of enrollment. The years of 1967 and 1968 appear similar, while the years 1969 and 1970 are more alike. These two groups differ from one another on retention rates.

We have examined only two variables, age and year of enrollment, as factors. Other variables—size of subscriber unit, number of children, age of youngest child—may be related to the probability of retention and may be used to obtain improved estimates of retention. The Kaiser Plan's administration must determine whether there is some change in the plan's operation that relates to a lower probability of retention among the youngest age group. The differences observed here may not be related to the plan's operation; they may also be the result of a downturn in Portland's economy, which would cause many of the newly employed (the younger group) to become unemployed and to lose or cancel their health insurance benefits.

SUMMARY

In this chapter we presented the calculation of a complex response variable: the probability of "surviving" a certain length of time. This function is complex, but we analyze it in the same fashion as the other functions considered in preceding chapters. The example developed in this chapter shows that follow-up life table analysis is not limited to clinical trial situations but also applies when we wish to track and compare experiences over time.

The key points covered in this chapter are:

- *The problem:* Measure the survival (retention) experience of HMO members; examine the relationship to age and year of enrollment.
- *Follow-up life table:* A method of analysis for comparing treatment intervention strategies.
- *WLS and follow-up life table:* Convert the data to a contingency table format; through multiple transformations form the probability of survival (retention).
- *Model specification:* A modular model provides an adequate fit to the data.

· *Results:* Younger and more recently joined groups had lower probabilities of retention.

This chapter completes our analysis of data. In the following chapter we survey some advanced applications of the WLS approach that complement those presented in Chapters 4 to 11.

EXERCISES

11.1.

The 5-year survival data shown in Table 11.3 were originally presented by Cutler and Ederer (1958) and were reanalyzed by Koch, Johnson, and Tolley (1972).

a. Construct the 4-year survival rate—that is, survived to the fourth year—for these data by using the GENCAT program.
b. What is the 5-year survival rate?

Table 11.3   Five-Year Survival Data for 126 Cases with
Localized Kidney Cancer

Years (y) After Diagnosis	Survived the Year	Died During Year	Withdrawn or Lost	Total Alive at Beginning of Year (y)
$0 \leq y < 1$	60	47	19	126
$1 \leq y < 2$	38	5	17	60
$2 \leq y < 3$	21	2	15	38
$3 \leq y < 4$	10	2	9	21
$4 \leq y < 5$	4	0	6	10

Source: Koch, Johnson, and Tolley (1972).

11.2.

The data in Table 11.4 from Potter (1966) are derived from the Princeton Fertility Study. The data represent the contraceptive experience of 495 native, once-married, white couples from seven large metropolitan areas of the United States. The period under study is the first 12 months directly following a second birth for couples who are using a contraceptive to postpone a desired third birth.

a. Calculate the 1-year failure rate for these data.
b. What advice would you give these couples regarding their chances of avoiding a third pregnancy?

Table 11.4   Contraceptive Experience of 495 White Couples

Months After Second Birth	Survived the Month	Contraceptive Failure	Deliberately Interrupted Contraception	Total Beginning the Month
1	492	2	1	495
2	490	2	0	492
3	487	2	1	490
4	483	3	1	487
5	471	9	3	483
6	461	8	2	471
7	448	12	1	461
8	431	16	1	448
9	418	11	2	431
10	402	14	2	418
11	384	11	7	402
12	371	11	2	384

11.3.

Consider the breast cancer data shown in Table 11.5, which were originally reported by Cutler and Myers (1967) and reanalyzed by Koch, Johnson, and Tolley (1972). Examine the relationship between the 5-year survival rate and these three factors:

1. Degree of skin fixation (S)
   a. None (S_0)
   b. Incomplete (S_1)
   c. Complete (S_2)
2. Node status (N)
   a. Clinically negative (N_0)
   b. Palpable (N_1)
3. Tumor size (T)
   a. 2 centimeters or less (T_0)
   b. More than 2 centimeters but less than 4 centimeters (T_1)
   c. More than 4 centimeters (T_2)

   a. Use the direct-input version of GENCAT and fit a saturated model incorporating the main and interaction effects of the three factors: degree of skin fixation (S), node status (N), and tumor size (T).

   b. Can the model be simplified? Fit a reduced model if it seems warranted.

   c. What conclusions would you draw from these data regarding the effects of the degree of skin fixation, node status, and tumor size on the 5-year survival rate?

**Table 11.5   Survival Rates for Recurrence of
Breast Cancer by Three Characteristics of the Cancer**

Category	Number of Cases	Five-Year Survival Rate	Estimated SE
$S_0N_0T_0$	195	0.88	0.024
$S_0N_0T_1$	226	0.77	0.028
$S_0N_0T_2$	96	0.62	0.050
$S_0N_1T_0$	72	0.78	0.049
$S_0N_1T_1$	89	0.67	0.050
$S_0N_1T_2$	53	0.49	0.069
$S_1N_0T_0$	41	0.95	0.034
$S_1N_0T_1$	114	0.74	0.042
$S_1N_0T_2$	78	0.51	0.057
$S_1N_1T_0$	24	0.63	0.099
$S_1N_1T_1$	55	0.58	0.066
$S_1N_1T_2$	59	0.57	0.065
$S_2N_0T_0$	15	0.93	0.069
$S_2N_0T_1$	30	0.67	0.086
$S_2N_0T_2$	26	0.38	0.095
$S_2N_1T_0$	7	0.71	0.171
$S_2N_1T_1$	15	0.47	0.129
$S_2N_1T_2$	38	0.39	0.079

APPENDIX: GENCAT INPUT

Estimation of F and V_F from Frequency Data in Table 11.1

The following listing is for the GENCAT control cards that estimate **F** and
V_F from the frequency data in Table 11.1.

```
     5     1     1              1      ESTIMATION FROM TABLE 11.1,PART I
     6      11                         (11F3.0)
 24  16  12  12   5   8   5            71
 23  27  20  11   5  12   5           111
  8  17   7   9   6  11   2            72
 41  28  14   8   3   2   5            74
 45  33  10  12  10   8   7           102
 18  13  10   5   3   5   3            67
     1       2    72     12     1
       1.    1.    1.    1.    1.    1.    1.    1.    1.    1.    1.    1.
 1.    1.    1.    1.    1.    1.    1.    1.    1.    1.    1.    1.
       1.    1.    1.    1.    1.    1.    1.    1.    1.    1.    1
       1.    1.    1.    1.    1.    1.    1.    1.    1.    1.    1
             1.    1.    1.    1.    1.    1.    1.    1.    1.
       1.    1.    1.    1.    1.    1.    1.    1.    1.
```

```
                     1.    1.    1.    1.    1.    1.    1.
               1.    1.    1.    1.    1.    1.    1.    1.
                     1.    1.    .5    1.    1.    1.
               1.    1.    1.    .5    1.    1.    1.
                           1.          .5    1.    1.
                     1.    1.          .5    1.    1.
      2                          1
      1     2     6     1     1
1.    -1.   1.    -1.   1.    -1.   1.    -1.   1.    -1.   1.    -1.
      3
      5     1     1           1           ESTIMATION FROM TABLE 11.1,PART II
      6    11                             (11F3.0)
 56  20  17   7   8  13   3          15  44
 61  33  14  13  11  12   8          43  78
 20  13   9   5   4   4   6          18  48
 43  24  11   9   8   1      29  20   7
 51  34  10  15  13          38  53  30
 16   8   7   8              25  24  23
      1     2    72    12     1
      1.    1.    1.    1.    1.    1.    1.    1.    1.    1.    1.    1.
1.    1.    1.    1.    1.    1.    1.    1.    1.    1.    1.    1.
            1.    1.    1.    1.    1.    1.    1.    1.    1.    1    1
      1.    1.    1.    1.    1.    1.    1.    1.    1.    1.    1
            1.    1.    1.    1.    1.    1.    1.    1.
            1.    1.    1.    1.    1.    1.    1.    1.
                  1.    1.    1.    1.    1.    1.    1.
            1.    1.    1.    1.    1.    1.    1.    1.
                  1.    1.    .5    1.    1.    1.
            1.    1.    1.    .5    1.    1.    1.
                        1.          .5    1.    1.
                  1.    1.          .5    1.    1.
      2                          1
      1     2     6     1     1
1.    -1.   1.    -1.   1.    -1.   1.    -1.   1.    -1.   1.    -1.
      3
```

Direct Input of Functions and Estimated Variance-Covariance Matrix

The following listing is for the GENCAT control cards that reproduce the analysis described in this chapter.

```
      5     3     1        1  CHAP 11: FOLLOW-UP LIFE TABLE ANALYSIS
     12    12     1           (12F6.0)
.49673.54206.56061.45143.48018.56452.33880.47253.56693.33989.48710.64865
      3                       (8F10.8)
.0016339   .00116    .0018661   .0014151   .0010996   .0019826   .0012241   .00091299
.0019332   .0018077   .0010584   .0020532
      7     1     6           (40F2.0)        AGE AND YEAR MAIN EFFECTS
 1  1  1  1  1  1  1  1  1  1  1
 1  1  1             -1-1-1
       1  1  1             -1-1-1
                1  1  1-1-1-1
 1  -1  1  -1  1  -1  1  -1
  1-1    1-1    1-1    1-1
      8     1     3           (80F1.0)        YEAR
 1
   1
     1
      8     1     2           (80F1.0)        AGE
   1
     1
```

```
    6                              REANALYSIS OF FOLLOW-UP LIFE TABLE
    7    1    5
1.   1.   1.   1.   1.   1.   1.   1.   1.   1.   1.   1.
-1.        1.
                  -1.        1.
                             -1.        1.
                                        -1.        1.
    8    1    1
     1.   -1.
    8    1    1
          1.   -1.
    8    1    1
               1.   -1.
    6                              REANALYSIS OF FOLLOW-UP LIFE TABLE
    7    1    3
1.   1.   1.   1.   1.   1.   1.   1.   1.   1.   1.   1.
-1.        1.   -1.        1.
                             -1.        1.   -1.        1.
    8    1    1                     (6F2.0)
 1-1
```

12
Selected WLS Literature

The preceding chapters have amply demonstrated the use of WLS methodology in public health and public affairs and shown the flexibility of the approach. In this chapter we describe some advanced applications and discuss recent theoretical developments which extend the range of problems suitable for treatment in the WLS framework. We have ignored clinical trials and experimental design, as well as biopharmaceutical, bioassay, and related biophysical problems, unless they have direct relevance. The following sections highlight a number of interesting problems to which researchers have applied WLS methodology. Although we do not cite non-WLS literature pertinent to these applications, we certainly recognize that alternative analytic approaches exist.

PAIRED COMPARISONS

Paired comparisons concern scaling techniques used to summarize individual behavior statistically by choosing paired alternatives selected from an available set of three or more objects or concepts. Such techniques are commonly applied to measurement and testing problems found in psychology, social psychology, and market research. One of the most common statistical models for paired-comparison data is that of Bradley and Terry (1952). Koch, Abernathy, and Imrey (1975) have used the WLS approach to estimate desired family size based on applying a Bradley–Terry model to paired comparisons obtained in the North Carolina Abortion Survey. Imrey, Johnson, and Koch (1976) develop multifactorial and multiresponse extensions and apply them to food preference data as well as to artificial data on policy alternatives in highway safety. Zimmerman and Rahlfs (1976) and Beaver (1977) provide a modification allowing for tied data.

OBSERVER AGREEMENT

Observer agreement is another problem common to measurement and testing. It arises whenever two or more "judges" are required to evaluate a subjective criterion. Landis and Koch (1977b) consider this general problem in the context of determining the agreement among physicians about patient diagnosis. For related discussions see Landis and Koch (1977a, 1977c), Corn, Landis and Flora (1977), and Frohardt-Lane, Landis, and Bruvold (1977).

SUPPLEMENTED MARGINS

The problem of *supplemented margins* arises whenever a sample is measured across a full set of variables while a subsample receives additional measurements on a subset of variables. Reinfurt (1970) uses data from a perinatal health survey to investigate whether there was an association between the Apgar score, a summary measure of the infant's health status at birth, and the serum bilirubin reading. Measurements on serum bilirubin are unnecessary for infants with low Apgar scores and normal complexions or for infants with advanced jaundice. Hence the serum bilirubin readings and Apgar readings are supplemented with additional Apgar readings alone for some of the premature births.

Lehnen and Koch (1974b) and Imrey and Koch (1972) apply this analytic technique to candidate preference data derived from 1968 presidential campaign surveys. In this application, a sample of respondents was given a lengthy questionnaire in a household survey conducted in September and was reinterviewed in October and November with a shortened questionnaire using telephone follow-ups. Koch, Imrey, and Reinfurt (1972) also consider applications to data on dissemination of cancer information and to trials of driver training techniques.

REPEATED MEASURES

Koch and others (1977) demonstrate how WLS methodology can be used to analyze multivariate categorical data obtained from repeated measurement studies. One example they consider involves comparison of the effects of three drugs in a two-period changeover design. Other examples of the analysis of repeated measures—also called *mixed models*—are provided by Koch and Reinfurt (1971), Lehnen and Koch (1974a), and Koch, Freeman, and Lehnen (1976).

PARTIAL ASSOCIATION

Landis, Heyman, and Koch (1978) summarize different measures of partial association in multiway tables. They also demonstrate the use of Functional Asymptotic Regression Methodology (FARM)—an extension of the WLS approach to small-sample problems discussed later in this chapter—to examine the relationship between atomic bomb radiation and the incidence of leukemia adjusted for age at exposure. Koch and Reinfurt (1974) also have considered extensions of the Cochran–Mantel–Haenszel procedure, a measure of average partial association, in the analysis of highway safety data.

COMPETING RISKS

The analysis of survival and follow-up life tables considered in Chapter 11 is a special case of the analysis of competing risks. Johnson and Koch (1978) discuss the problem of analyzing competing risks in a study of duodenal ulcer concerned with death or recurrence by treating incomplete information due to follow-up loss or reoperation as a competing risk. Another type of extension concerning survival data is found in Freeman, Freeman, and Koch (1978).

MULTIPLE-RECORD SYSTEMS

Record systems have varying degrees of coverage for the events they are supposed to record. The vital statistics system covers almost 100 percent of births and deaths. Disease-specific registries, business mailing lists, reports about crime gathered from calls to police and victim surveys, and traffic accident summaries are examples of systems that have incomplete coverage of their target populations. Many organizations use multiple sources or lists in an effort to estimate the true total number of events, people, crimes, and accidents in the population. Koch, El-Khorazaty, and Lewis (1976) demonstrate the use of multiple-record systems to estimate the total number of children in Massachusetts born between 1 January, 1955 and 31 December, 1959; who are still alive on 31 December, 1966; and who received a positive diagnosis of Down's syndrome by use of records from hospitals, obstetric units, schools, and the Massachusetts Department of Health and Mental Health. El-Khorazaty and others (1977) review current research pertaining to estimation from multiple-record systems in the context of log-linear models.

COMPLEX SAMPLE SURVEYS

Data from many national surveys—for example, the National Health Interview Survey, the Current Population Survey, and the National Crime Survey—result from complex sampling strategies, not from simple random samples. Complex samples require that estimates of the variance-covariance matrix be adjusted to reflect the characteristics of the design. Koch, Freeman, and Freeman (1975); Freeman and others (1976a, 1976b); Koch and others (1976); Koch, Gillings, and Stokes (1980); and Johnson and Koch (1970) discuss estimation problems common to complex surveys and incorporate appropriate estimates of the variance-covariance matrix into the analysis of data from complex sample surveys. Imrey, Sobel, and Francis (1979) consider the true variance-covariance matrices of contingency tables derived from complex sample surveys and the parameter estimates stemming from applications of WLS analysis. These papers also discuss extensions of standard sampling theory results for proportions based on stratified or cluster random sampling.

COMPLEX FUNCTIONS

Forthofer and Koch (1973) extend the WLS approach beyond simple linear and logarithmic functions by providing a theoretical basis for the use of exponential functions. Before this development, applications were confined to a simple linear or a logarithmic-linear sequence of transformations. A further application of this work involves the use of these three transformations in any combination and any order.

FUNCTIONAL ASYMPTOTIC REGRESSION METHODOLOGY (FARM)

Koch and others (1976) discuss the use of the WLS approach to fit log-linear and logistic models when there are subpopulations with fewer than 10 observations. In this situation the estimation of population statistics by a WLS procedure is not suitable because the asymptotic features of WLS are not strictly applicable to small samples. Maximum likelihood estimation has an advantage over WLS in this case, since the maximum likelihood estimates are based on marginal totals, not on the individual cell frequencies. The FARM approach uses the maximum likelihood method to estimate the parameters of a preliminary model containing main effects and lower-order interactions and the corresponding cell probabilities of the contingency table. After these estimates have been obtained, the weighted least squares approach is applied to functions of interest based on the first-stage estimates to assess the plausibility of more parsimonious models. The use of FARM in substantive examples has been referred to elsewhere in this chapter.

VARIABLE SELECTION

An analyst can often identify more candidate variables for use as predictors than can be conveniently included in the analysis. Limitations resulting from sample size, the analyst's time, and computer resources require that the problem be simplified. Higgins and Koch (1977) suggest a method for selecting a promising subset of explanatory variables and thereby simplifying the analysis.

Their approach is based on the Pearson chi-square and Cochran-Mantel-Haenszel test statistics that guide the analyst in selecting a subset of predictor variables that will maximize explained variation. This variable selection procedure, which is analogous to forward-stepwise regression selection, is applied to an occupational health problem (Higgins and Koch, 1977) and to criminal justice data (Clarke and Koch, 1976).

COMPARISONS TO PARAMETRIC REGRESSION

Several papers have compared the WLS approach to applications of parametric regression with dichotomous dependent variables. Kuechler (1980) compares parametric regression and WLS for saturated models; Lehnen and Koch (1973) discuss an application for reduced models.

PHILOSOPHY OF DATA ANALYSIS

Three publications by Koch and others discuss important issues affecting the method of analysis, the degree of rigor, and the scope of generalization applied to a set of data. Koch, Gillings, and Stokes (1980) present a general discussion of these issues. Koch and Stokes (1981) and Koch and Bhapkar (1981) review issues pertaining to the use of chi-square tests.

Appendixes

Appendix A:
Matrix Notation

Matrix notation is a shorthand notation that is useful in applications such as statistics and physics. Matrix notation is compact, elegant, and more convenient than scalars in complex situations. The advent of the computer has played an important role in the new popularity of matrices, as matrix calculations are unwieldy for hand calculations but can be handled swiftly by a computer.

This appendix offers a brief introduction to the essential matrix theory needed for this book. A more thorough treatment of matrices is given in books by Strang (1976), Searle (1966), and Graybill (1969).

MATRIX DEFINITIONS

Definition: A *matrix* is a rectangular array of symbols. It is characterized by its physical dimensions—that is, the number of rows and columns in it.

Consider the matrix **A**, which has three rows and four columns. Matrix **A** can be written as:

$$\mathop{\mathbf{A}}_{(3 \times 4)} = \begin{bmatrix} a_{11} & a_{12} & a_{13} & a_{14} \\ a_{21} & a_{22} & a_{23} & a_{24} \\ a_{31} & a_{32} & a_{33} & a_{34} \end{bmatrix}$$

Note that the subscripts on the elements of **A** give the row and column position of the element. The element a_{ij} is in the ith row and jth column of **A**. The numbers under **A** indicate the dimension of the matrix: the number of rows is given first and the number of columns second. Additional examples of matrices are

$$\underset{(1 \times 1)}{\mathbf{B}} = [3] \qquad \underset{(3 \times 1)}{\mathbf{c}} = \begin{bmatrix} -a \\ b \\ 0 \end{bmatrix} \qquad \underset{(1 \times 2)}{\mathbf{d}} = [-4 \quad 0.6]$$

$$\underset{(2 \times 3)}{\mathbf{E}} = \begin{bmatrix} 1/8 & 2 & 3 \\ -4 & 6.1 & 3 \end{bmatrix} \qquad \underset{(2 \times 2)}{\mathbf{F}} = \begin{bmatrix} f_{11} & f_{12} \\ f_{21} & f_{22} \end{bmatrix}$$

Matrix **B** has one row and one column; that is, it is a *scalar*. Matrix **c** has only one column and may also be referred to as a *column vector*. Matrix **d** has only one row and may be called a *row vector*. Both vectors and matrices are set in boldface type, and vectors are often set in lower case.

To refer to a particular element in a matrix, we cite its row and column position. For example, the element in the second row and second column of **E** is e_{22}, and $e_{22} = 6.1$. Note that in the definition of a matrix the elements of the matrix are *symbols*. In this book all the symbols have real numerical values.

The matrix **F** is a *square matrix* since it has the same number of rows as columns.

Definition: The square matrix $\underset{(n \times n)}{\mathbf{A}}$ is a *diagonal matrix* if $a_{ij} = 0$ for i and $j = 1, 2, \ldots, n$ and $i \neq j$. Example:

$$\underset{(3 \times 3)}{\mathbf{A}} = \begin{bmatrix} 1 & 0 & 0 \\ 0 & -1 & 0 \\ 0 & 0 & 2.5 \end{bmatrix}$$

is a diagonal matrix since all the elements off the main diagonal (running from upper left to lower right) are zero. An especially useful type of diagonal matrix is the identity matrix.

Definition: An $n \times n$ diagonal matrix with all the diagonal elements equal to 1 is called the *identity matrix* and denoted by $\underset{(n \times n)}{\mathbf{I}}$ or $\mathbf{I}_n$.

The diagonal matrix can be extended. If the elements of a diagonal matrix are matrices rather than scalars, and all the matrices off the main diagonal contain nothing but zeros, the matrix is called a *block-diagonal matrix*. Example:

$$\mathbf{A}_{(5 \times 5)} = \begin{bmatrix} 1 & 2 & 0 & 0 & 0 \\ 3 & 4 & 0 & 0 & 0 \\ 0 & 0 & -1 & 2 & 0 \\ 0 & 0 & 3 & -2 & 1.5 \\ 0 & 0 & 6 & 2 & 1 \end{bmatrix}$$

or

$$\mathbf{A} = \begin{bmatrix} \mathbf{B} & \mathbf{C} \\ {\scriptstyle (2 \times 2)} & {\scriptstyle (2 \times 3)} \\ \mathbf{D} & \mathbf{E} \\ {\scriptstyle (3 \times 2)} & {\scriptstyle (3 \times 3)} \end{bmatrix}$$

and the elements in **C** and **D** are all zeros. A block-diagonal matrix in which all the submatrices contain one row and one column is a diagonal matrix.

We now consider some further properties and operations that constitute matrix arithmetic.

Definition:　Matrices **A** and **B** are said to be *equal* if and only if (1) **A** and **B** have the same dimensions and (2) $a_{ij} = b_{ij}$ for all values of i and j for which a_{ij} is defined.

Definition:　The matrix $\mathbf{B} = \mathbf{A}'$ (or $\mathbf{A}^T$) is the *transpose* of **A** if and only if $b_{ij} = a_{ji}$ for all values of i and j for which a_{ij} is defined. Example:

$$\mathbf{A}_{(2 \times 3)} = \begin{bmatrix} 1 & 2 & 3 \\ 4 & 5 & 6 \end{bmatrix} \qquad \mathbf{A}'_{(3 \times 2)} = \begin{bmatrix} 1 & 4 \\ 2 & 5 \\ 3 & 6 \end{bmatrix}$$

As you can see, the notion of rows and columns has been interchanged—that is, the elements that made up the first row of **A** now constitute the first column of **A'**. The elements that formed the jth column of **A** are now the jth row of **A'**.

Definition:　The matrix **A** is a *symmetric matrix* if $\mathbf{A} = \mathbf{A}'$. Only square matrices may be symmetric.

MATRIX ARITHMETIC

Definition:　If matrices **A** and **B** have the same dimensions, then *matrix addition* of **A** and **B** is the sum of the elements in the i,jth position in **A** and **B**—that is, $a_{ij} + b_{ij}$, for all values of i and j for which a_{ij} is defined. Example:

$$\mathop{\mathbf{A}}_{(2\times 3)} = \begin{bmatrix} 1 & 2 & 3 \\ 3.5 & -2 & 6 \end{bmatrix} \quad \mathop{\mathbf{B}}_{(2\times 3)} = \begin{bmatrix} -1 & 0 & 2 \\ -3.5 & -2 & 4 \end{bmatrix}$$

$$\mathop{\mathbf{C}}_{(2\times 3)} = \mathbf{A} + \mathbf{B} = \begin{bmatrix} (1 + -1) & (2 + 0) & (3 + 2) \\ (3.5 + -3.5) & (-2 + -2) & (6 + 4) \end{bmatrix}$$

$$= \begin{bmatrix} 0 & 2 & 5 \\ 0 & -4 & 10 \end{bmatrix}$$

If

$$\mathop{\mathbf{D}}_{(3\times 2)} = \begin{bmatrix} 1 & a \\ 2 & 6 \\ 7 & 4 \end{bmatrix}$$

$\mathbf{A} + \mathbf{D}$, $\mathbf{B} + \mathbf{D}$, and $\mathbf{C} + \mathbf{D}$ are not defined since they do not have the same dimensions.

Definition: If matrices $\mathbf{A}$ and $\mathbf{B}$ have the same dimensions, then *matrix subtraction* of $\mathbf{B}$ from $\mathbf{A}$ is the subtraction of the element in the i,jth position in $\mathbf{B}$ from the element in the i,jth position in $\mathbf{A}$—that is, $a_{ij} - b_{ij}$, for all values of i and j for which a_{ij} is defined.

Definition: *Scalar multiplication*, the product of a scalar c and a matrix $\mathbf{A}$, is defined as the product of c and a_{ij} for all values of i and j for which a_{ij} is defined. Example:

$$\mathop{\mathbf{A}}_{(2\times 3)} = \begin{bmatrix} 1 & 2 & 3 \\ 3.5 & -2 & 6 \end{bmatrix} \quad c = 3$$

$$\mathop{\mathbf{B}}_{(2\times 3)} = c\mathbf{A} = \mathbf{A}c = \begin{bmatrix} 3 & 6 & 9 \\ 10.5 & -6 & 18 \end{bmatrix}$$

Definition: If the number of columns in $\mathop{\mathbf{A}}_{(m\times n)}$ equals the number of rows in $\mathop{\mathbf{B}}_{(n\times p)}$, then the *matrix multiplication* of $\mathbf{A}$ and $\mathbf{B}$—that is, $\mathbf{C} = \mathbf{A} \times \mathbf{B}$—is defined by

$$c_{ij} = \sum_{k=1}^{n} a_{ik}b_{kj}$$

for $i = 1, 2, \ldots, m$ and $j = 1, 2, \ldots, p$.

This definition tells us that the product matrix $\mathbf{C}$ has as many rows as there are in $\mathbf{A}$ and as many columns as in $\mathbf{B}$. It also says that $\mathbf{D} = \mathop{\mathbf{B}}_{(n\times p)} \times \mathop{\mathbf{A}}_{(m\times n)}$ is not defined unless $p = m$. Immediately we see that $\mathbf{A} \times \mathbf{B} \neq \mathbf{B} \times \mathbf{A}$ because in some cases one side of this equation may not be defined when the other side is.

Let us apply this definition in an example:

$$\underset{(2 \times 2)}{\mathbf{C}} = \underset{(2 \times 3)}{\mathbf{A}} \times \underset{(3 \times 2)}{\mathbf{B}}$$

where

$$\underset{(2 \times 3)}{\mathbf{A}} = \begin{bmatrix} 1 & 2 & 3 \\ 3.5 & -2 & 6 \end{bmatrix}$$

and

$$\underset{(3 \times 2)}{\mathbf{B}} = \begin{bmatrix} 1 & 2 \\ 0 & 3 \\ 4 & 0 \end{bmatrix}$$

$$c_{11} = \sum_{k=1}^{3} a_{1k}b_{k1}$$

$$= a_{11}b_{11} + a_{12}b_{21} + a_{13}b_{31}$$
$$= (1 \cdot 1) + (2 \cdot 0) + (3 \cdot 4)$$
$$= 13$$

$$c_{12} = \sum_{k=1}^{3} a_{1k}b_{k2}$$

$$= a_{11}b_{12} + a_{12}b_{22} + a_{13}b_{32}$$
$$= (1 \cdot 2) + (2 \cdot 3) + (3 \cdot 0)$$
$$= 8$$

$$c_{21} = \sum_{k=1}^{3} a_{2k}b_{k1}$$

$$= a_{21}b_{11} + a_{22}b_{21} + a_{23}b_{31}$$
$$= 3.5 \cdot 1 + (-2) \cdot (0) + 6 \cdot 4$$
$$= 27.5$$

$$c_{22} = \sum_{k=1}^{3} a_{2k}b_{k2}$$

$$= a_{21}b_{12} + a_{22}b_{22} + a_{23}b_{32}$$
$$= 3.5 \cdot 2 + (-2) \cdot (3) + 6 \cdot 0$$
$$= 1$$

$$\underset{(2 \times 2)}{\mathbf{C}} = \begin{bmatrix} 13 & 8 \\ 27.5 & 1 \end{bmatrix}$$

Careful inspection of the terms used in the calculation of c_{11} shows that it is formed by taking the first row of **A** times the first column of **B** and summing each of the products:

$$[1 \quad 2 \quad 3]\begin{bmatrix} 1 \\ 0 \\ 4 \end{bmatrix} = (1 \cdot 1) + (2 \cdot 0) + (3 \cdot 4) = 13$$

In the same way c_{12} is the result of the first row of **A** times the second column of **B**:

$$[1 \quad 2 \quad 3]\begin{bmatrix} 2 \\ 3 \\ 0 \end{bmatrix} = (1 \cdot 2) + (2 \cdot 3) + (3 \cdot 0) = 8$$

In general c_{ij} is found by multiplying the ith row of **A** times the jth column of **B**.

Matrix multiplication plays an important role in the specification of models and testing of hypotheses. Since it is essential that it be understood, we consider one more example of matrix multiplication:

$$\underset{(3 \times 3)}{\mathbf{I}} = \begin{bmatrix} 1 & 0 & 0 \\ 0 & 1 & 0 \\ 0 & 0 & 1 \end{bmatrix} \qquad \underset{(3 \times 2)}{\mathbf{D}} = \begin{bmatrix} 4 & 7 \\ 0 & -1 \\ 6 & 3 \end{bmatrix}$$

$$\underset{(3 \times 2)}{\mathbf{E}} = \mathbf{I} \times \mathbf{D}$$

$$e_{11} = \text{row 1 of } \mathbf{I} \times \text{ column 1 of } \mathbf{D}$$

$$= [1 \quad 0 \quad 0]\begin{bmatrix} 4 \\ 0 \\ 6 \end{bmatrix} = 4$$

$$e_{12} = \text{row 1 of } \mathbf{I} \times \text{ column 2 of } \mathbf{D}$$

$$= [1 \quad 0 \quad 0]\begin{bmatrix} 7 \\ -1 \\ 3 \end{bmatrix} = 7$$

$$e_{21} = \text{row 2 of } \mathbf{I} \times \text{ column 1 of } \mathbf{D}$$

$$= [0 \quad 1 \quad 0]\begin{bmatrix} 4 \\ 0 \\ 6 \end{bmatrix} = 0$$

e_{22} = row 2 of **I** × column 2 of **D**

$$= \begin{bmatrix} 0 & 1 & 0 \end{bmatrix} \begin{bmatrix} 7 \\ -1 \\ 3 \end{bmatrix} = -1$$

e_{31} = row 3 of **I** × column 1 of **D**

$$= \begin{bmatrix} 0 & 0 & 1 \end{bmatrix} \begin{bmatrix} 4 \\ 0 \\ 6 \end{bmatrix} = 6$$

e_{32} = row 3 of **I** × column 2 of **D**

$$= \begin{bmatrix} 0 & 0 & 1 \end{bmatrix} \begin{bmatrix} 7 \\ -1 \\ 3 \end{bmatrix} = 3$$

$$\mathbf{E} = \begin{bmatrix} 4 & 7 \\ 0 & -1 \\ 6 & 3 \end{bmatrix} = \mathbf{I} \times \mathbf{D} = \mathbf{D}$$

This example illustrates an application of the identity matrix. Note that $\mathbf{I}_{(n \times n)}$ plays the same role in matrix multiplication as the number 1 in scalar multiplication.

The transpose of the product of two matrices **A** and **B** is equal to **B′A′**— that is, $(\mathbf{AB})' = \mathbf{B'A'}$. The following example demonstrates this property of matrix multiplication. Let

$$\underset{(2 \times 3)}{\mathbf{A}} = \begin{bmatrix} 1 & 2 & 3 \\ 0 & 1 & 0 \end{bmatrix} \quad \text{and} \quad \underset{(3 \times 3)}{\mathbf{B}} = \begin{bmatrix} 0 & -1 & 1 \\ -1 & 2 & 4 \\ 0 & 1 & 3 \end{bmatrix}$$

$$\underset{(2 \times 3)}{\mathbf{C}} = \mathbf{A} \times \mathbf{B} = \begin{bmatrix} -2 & 6 & 18 \\ -1 & 2 & 4 \end{bmatrix}$$

$$\underset{(3 \times 2)}{\mathbf{C'}} = \begin{bmatrix} -2 & -1 \\ 6 & 2 \\ 18 & 4 \end{bmatrix}$$

$$\underset{(3 \times 3)}{\mathbf{B'}} \times \underset{(3 \times 2)}{\mathbf{A'}} = \begin{bmatrix} 0 & -1 & 0 \\ -1 & 2 & 1 \\ 1 & 4 & 3 \end{bmatrix} \times \begin{bmatrix} 1 & 0 \\ 2 & 1 \\ 3 & 0 \end{bmatrix}$$

$$= \begin{bmatrix} -2 & -1 \\ 6 & 2 \\ 18 & 4 \end{bmatrix} = \mathbf{C'}$$

THE INVERSE MATRIX

Division is an arithmetic operation in scalars; the matrix analog is *inversion*.

Definition: **B** is the *inverse matrix* of **A** (denoted by A^{-1}) if and only if $B \times A = A \times B = I$.

Note that this definition is confined to the special case where **A** and **B** are square. This is implied by $B \times A = A \times B = I$. We do not consider inverses for nonsquare matrices in this book.

If a matrix has an inverse matrix, the matrix is said to be *nonsingular*. If the inverse matrix does not exist, the matrix is said to be *singular*. A square matrix is singular if one of its rows (columns) is a linear combination of its other rows (columns).

We now consider an example of a singular matrix. Let

$$C = \begin{bmatrix} 1 & 2 & 3 \\ 2 & -1 & 0 \\ 0 & 5 & 6 \end{bmatrix}$$

The third row of **C** is formed by subtracting the second row from twice the first row. Hence the third row of **C** is a linear combination of the first two rows of **C**, and this implies that C^{-1} does not exist.

To find the inverse matrix, we can solve the equation $A \times B = I$ for the elements of **B** or **A**. To demonstrate this let us work with the matrix **A**, which has two rows and two columns. We assume that we know the elements of **A** and wish to find its inverse matrix **B**. We have

$$\begin{bmatrix} a_{11} & a_{12} \\ a_{21} & a_{22} \end{bmatrix} \times \begin{bmatrix} b_{11} & b_{12} \\ b_{21} & b_{22} \end{bmatrix} = \begin{bmatrix} 1 & 0 \\ 0 & 1 \end{bmatrix}$$

and we wish to solve for the b_{ij}. Matrix multiplication yields four equations in four unknowns, the b_{ij}:

$$a_{11}b_{11} + a_{12}b_{21} = 1 \tag{A.1}$$

$$a_{11}b_{12} + a_{12}b_{22} = 0 \tag{A.2}$$

$$a_{21}b_{11} + a_{22}b_{21} = 0 \tag{A.3}$$

$$a_{21}b_{12} + a_{22}b_{22} = 1 \tag{A.4}$$

Solve Equation (A.1) for b_{11} in terms of b_{21} and the a_{ij}'s:

$$b_{11} = \frac{1 - a_{12}b_{21}}{a_{11}} \tag{A.5}$$

Substitute this expression into Equation (A.3) for b_{11} and solve for b_{21}:

$$a_{21} \frac{1 - a_{12}b_{21}}{a_{11}} + a_{22}b_{21} = 0$$

$$\frac{a_{21}}{a_{11}} - \frac{a_{21}a_{12}}{a_{11}} b_{21} + a_{22}b_{21} = 0$$

$$b_{21} \frac{a_{11}a_{22} - a_{21}a_{12}}{a_{11}} = -\frac{a_{21}}{a_{11}}$$

$$b_{21} = \frac{-a_{21}}{a_{11}a_{22} - a_{21}a_{12}} \tag{A.6}$$

Substitution of this value into Equation (A.5) for b_{11} gives

$$b_{11} = \frac{1 - a_{12}\left(\dfrac{-a_{21}}{a_{11}a_{22} - a_{21}a_{12}}\right)}{a_{11}}$$

$$= \frac{a_{11}a_{22} - a_{21}a_{12} + a_{21}a_{12}}{a_{11}(a_{11}a_{22} - a_{21}a_{12})}$$

$$= \frac{a_{22}}{a_{11}a_{22} - a_{21}a_{12}} \tag{A.7}$$

Equations (A.2) and (A.4) similarly yield

$$b_{12} = \frac{-a_{12}}{a_{11}a_{22} - a_{21}a_{12}} \quad \text{and} \quad b_{22} = \frac{a_{11}}{a_{11}a_{22} - a_{21}a_{12}} \tag{A.8}$$

Hence

$$\mathbf{B} \text{ (or } \mathbf{A}^{-1}) = \frac{1}{a_{11}a_{22} - a_{21}a_{12}} \times \begin{bmatrix} a_{22} & -a_{12} \\ -a_{21} & a_{11} \end{bmatrix} \tag{A.9}$$

Consider the following numerical example:

$$\underset{(2 \times 2)}{\mathbf{A}} = \begin{bmatrix} 1 & -1 \\ 2 & 5 \end{bmatrix}$$

$$\mathbf{B} = \mathbf{A}^{-1} = \frac{1}{1 \cdot 5 - 2(-1)} \begin{bmatrix} 5 & 1 \\ -2 & 1 \end{bmatrix}$$

$$= \frac{1}{7} \begin{bmatrix} 5 & 1 \\ -2 & 1 \end{bmatrix} \quad \text{or} \quad \begin{bmatrix} \frac{5}{7} & \frac{1}{7} \\ -\frac{2}{7} & \frac{1}{7} \end{bmatrix}$$

To verify that $\mathbf{A} \times \mathbf{A}^{-1} = \mathbf{I} = \mathbf{A}^{-1} \times \mathbf{A}$, perform the indicated multiplication:

$$\begin{bmatrix} 1 & -1 \\ 2 & 5 \end{bmatrix} \times \begin{bmatrix} \frac{5}{7} & \frac{1}{7} \\ -\frac{2}{7} & \frac{1}{7} \end{bmatrix} = \begin{bmatrix} 1 & 0 \\ 0 & 1 \end{bmatrix}$$

$$\begin{bmatrix} \frac{5}{7} & \frac{1}{7} \\ -\frac{2}{7} & \frac{1}{7} \end{bmatrix} \times \begin{bmatrix} 1 & -1 \\ 2 & 5 \end{bmatrix} = \begin{bmatrix} 1 & 0 \\ 0 & 1 \end{bmatrix}$$

We will not worry about finding the inverse of larger matrices by hand since matrix inverse routines have been programmed for the computer.

A careful review of the previous sections will provide sufficient command of the matrix theory needed for this book. The following section shows an application of matrix theory to solving equations.

SYSTEM OF LINEAR EQUATIONS— SCALAR PRESENTATION

One problem encountered frequently in mathematics is solving a system of linear equations. That is exactly what we did when we found the elements that defined the inverse of the 2 × 2 matrix. A scalar representation of this situation is

$$a_{11}x_1 + a_{12}x_2 + \cdots + a_{1p}x_p = b_1$$
$$a_{21}x_1 + a_{22}x_2 + \cdots + a_{2p}x_p = b_2$$
$$\vdots \qquad \vdots \qquad\qquad \vdots \qquad \vdots$$
$$a_{p1}x_1 + a_{p2}x_2 + \cdots + a_{pp}x_p = b_p$$

In this situation the a_{ij} are known coefficients, the b_i are known constants, and the x_i are the unknown variables. There are a number of ways of solving simultaneous linear equations. We consider a specific example to demonstrate the *substitution method*:

$$3x_1 + 2x_2 + 3x_3 = 10 \qquad\qquad \text{(A.10)}$$
$$1x_1 + 0x_2 - 1x_3 = 6 \qquad\qquad \text{(A.11)}$$
$$0x_1 + 3x_2 + 1x_3 = 9 \qquad\qquad \text{(A.12)}$$

The substitution method involves solving one equation for one variable in terms of the other variables and substituting that value into the other equations. The next step is a repetition of the same process with the reduced set of equations.

Let us solve Equation (A.10) for x_1 in terms of x_2 and x_3:

$$3x_1 + 2x_2 + 3x_3 = 10$$
$$3x_1 = 10 - 2x_2 - 3x_3$$

or

$$x_1 = \frac{10 - 2x_2 - 3x_3}{3} \tag{A.13}$$

Substitute this value of x_1 into Equations (A.11) and (A.12):

$$1\frac{10 - 2x_2 - 3x_3}{3} + 0x_2 - 1x_3 = 6$$

$$0\frac{10 - 2x_2 - 3x_3}{3} + 3x_2 + 1x_3 = 9$$

or

$$\tfrac{10}{3} - \tfrac{2}{3}x_2 - 1x_3 + 0x_2 - 1x_3 = 6$$
$$3x_2 + 1x_3 = 9$$

or

$$-\tfrac{2}{3}x_2 - 2x_3 = \tfrac{8}{3} \tag{A.14}$$
$$3x_2 + 1x_3 = 9 \tag{A.15}$$

This substitution process has taken us from three equations in three unknowns to two equations in two unknowns. We next repeat the process by solving Equation (A.14) for x_2 in terms of x_3:

$$-\tfrac{2}{3}x_2 - 2x_3 = \tfrac{8}{3}$$

or

$$-\tfrac{2}{3}x_2 = \tfrac{8}{3} + 2x_3$$

or

$$x_2 = -4 - 3x_3 \tag{A.16}$$

Substitute this value of x_2 into Equation (A.15):

$$3(-4 - 3x_3) + 1x_3 = 9$$

or

$$-12 - 9x_3 + 1x_3 = 9$$

or

$$-8x_3 = 21$$

or

$$x_3 = -\tfrac{21}{8} \tag{A.17}$$

To find the value of x_2, substitute x_3 into Equation (A.16). This yields

$$
\begin{aligned}
x_2 &= -4 - 3x_3 = -4 - 3 \cdot -\tfrac{21}{8} \\
&= -\tfrac{32}{8} + \tfrac{63}{8} \\
&= \tfrac{31}{8}
\end{aligned} \tag{A.18}
$$

The solution for x_1 is found by substituting the values of x_2 and x_3 into Equation (A.13):

$$
\begin{aligned}
x_1 &= \frac{10 - 2x_2 - 3x_3}{3} \\
&= \frac{10 - 2(\tfrac{31}{8}) - 3(-\tfrac{21}{8})}{3} \\
&= \frac{10 + \tfrac{1}{8}}{3} = \tfrac{81}{24} \\
&= \tfrac{27}{8}
\end{aligned} \tag{A.19}
$$

These values can be verified by substituting them into the original set of three equations:

$$3(\tfrac{27}{8}) + 2(\tfrac{31}{8}) + 3(-\tfrac{21}{8}) = (\tfrac{80}{8}) = 10$$
$$1(\tfrac{27}{8}) + 0(\tfrac{31}{8}) - 1(-\tfrac{21}{8}) = (\tfrac{48}{8}) = 6$$
$$0(\tfrac{27}{8}) + 3(\tfrac{31}{8}) + 1(-\tfrac{21}{8}) = (\tfrac{72}{8}) = 9$$

Therefore the solution is correct. Now let us approach the same problem by using matrices.

SYSTEM OF LINEAR EQUATIONS— MATRIX PRESENTATION

Note that the system of simultaneous linear equations

$$a_{11}x_1 + a_{12}x_2 + \cdots + a_{1p}x_p = b_1$$

$$a_{21}x_1 + a_{22}x_2 + \cdots + a_{2p}x_p = b_2$$

$$\vdots \qquad \vdots \qquad \qquad \vdots \quad \vdots$$

$$a_{p1}x_1 + a_{p2}x_2 + \cdots + a_{pp}x_p = b_p$$

may be written in matrix notation as

$$\underset{(p \times p)}{\mathbf{A}} \times \underset{(p \times 1)}{\mathbf{x}} = \underset{(p \times 1)}{\mathbf{b}} \qquad\qquad (A.20)$$

where

$$\mathbf{A} = \begin{bmatrix} a_{11} & a_{12} & \cdots & a_{1p} \\ a_{21} & a_{22} & \cdots & a_{2p} \\ \vdots & \vdots & & \vdots \\ a_{p1} & a_{p2} & \cdots & a_{pp} \end{bmatrix} \qquad \mathbf{x} = \begin{bmatrix} x_1 \\ x_2 \\ \vdots \\ x_p \end{bmatrix} \qquad \mathbf{b} = \begin{bmatrix} b_1 \\ b_2 \\ \vdots \\ b_p \end{bmatrix}$$

To see this, note that taking the ith row of $\mathbf{A}$ times $\mathbf{x}$ and setting it equal to the ith row of $\mathbf{b}$ yields

$$a_{i1}x_1 + a_{i2}x_2 + \cdots + a_{ip}x_p = b_p$$

Hence the matrix formulation produces the same equations as given previously. To solve the matrix equation (A.20) for $\mathbf{x}$, multiply both sides by $\mathbf{A}^{-1}$:

$$\mathbf{A}^{-1}\mathbf{A}\mathbf{x} = \mathbf{A}^{-1}\mathbf{b} \qquad\qquad (A.21)$$

which yields

$$\mathbf{I}\mathbf{x} = \mathbf{A}^{-1}\mathbf{b} \qquad\qquad (A.22)$$

and

$$\mathbf{x} = \mathbf{A}^{-1}\mathbf{b} \qquad\qquad (A.23)$$

The solution of a system of simultaneous linear equations involves finding the inverse of the matrix of coefficients. This relationship holds no matter how large p (the number of equations and variables) becomes.

Consider the numerical example introduced in Equations (A.10) to (A.12):

$$3x_1 + 2x_2 + 3x_3 = 10$$
$$1x_1 + 0x_2 - 1x_3 = 6$$
$$0x_1 + 3x_2 + 1x_3 = 9$$

$$\mathbf{A} = \begin{bmatrix} 3 & 2 & 3 \\ 1 & 0 & -1 \\ 0 & 3 & 1 \end{bmatrix} \quad \mathbf{x} = \begin{bmatrix} x_1 \\ x_2 \\ x_3 \end{bmatrix} \quad \mathbf{b} = \begin{bmatrix} 10 \\ 6 \\ 9 \end{bmatrix}$$

The inverse for the 3×3 matrix $\mathbf{A}$ is

$$\mathbf{A}^{-1} = \frac{1}{16} \begin{bmatrix} 3 & 7 & -2 \\ -1 & 3 & 6 \\ 3 & -9 & -2 \end{bmatrix} \qquad (A.24)$$

We can verify that this is the inverse of $\mathbf{A}$ by multiplying $\mathbf{A} \times \mathbf{A}^{-1}$ and $\mathbf{A}^{-1} \times \mathbf{A}$. Both these products yield the identity matrix. Substituting $\mathbf{A}^{-1}$ into Equation (A.23) gives

$$\mathbf{x} = \mathbf{A}^{-1}\mathbf{b} = \frac{1}{16} \begin{bmatrix} 3 & 7 & -2 \\ -1 & 3 & 6 \\ 3 & -9 & -2 \end{bmatrix} \cdot \begin{bmatrix} 10 \\ 6 \\ 9 \end{bmatrix}$$

$$= \begin{bmatrix} \frac{54}{16} \\ \frac{62}{16} \\ -\frac{42}{16} \end{bmatrix} = \begin{bmatrix} \frac{27}{8} \\ \frac{31}{8} \\ -\frac{21}{8} \end{bmatrix}$$

Note that these are exactly the same values given in Equations (A.17) to (A.19), and hence the scalar and matrix solutions are the same.

TWO TRANSFORMATIONS: e AND THE NATURAL LOGARITHMS

Transformations of data sometimes make it easier for the investigator to determine the relationship between sets of variables. Two transformations of data that have proved useful are the *exponential* and *natural logarithmic* transformations. Both are based on the numerical constant e, which is approximately

Table A.1   Relationship of x to e^x
for Selected Values

x	e^x
−10.0	$e^{-10} = 1/e^{10} = $ 0.00005
−5.0	$e^{-5} = 1/e^5 = $ 0.00674
−1.0	$e^{-1} = 1/e^1 = $ 0.36788
0.0	$e^0 \qquad\quad = $ 1.00000
0.5	$e^{0.5} = \sqrt{e} = $ 1.64872
1.0	$e^1 \quad = e \;\; = $ 2.71828
1.5	$e^{1.5} \qquad\quad = $ 4.48169
2.0	$e^2 \qquad\quad = $ 7.38906
3.0	$e^3 \qquad\quad = $ 20.08554
4.0	$e^4 \qquad\quad = $ 54.59815

Table A.2   Relationship of x
to Natural Logarithm of x for
Selected Values

x	$\ln(x)$
0.00005	−10.0
0.00674	−5.0
0.36788	−1.0
1.00000	0.0
1.64872	0.5
2.71828	1.0
4.48169	1.5
7.38906	2.0
20.08554	3.0
54.59815	4.0

equal to 2.71828. The exponential transformation of the variable x is written as $\exp(x)$ or e^x; the natural logarithmic transformation of x is written as $\log_e(x)$ or $\ln(x)$. Tables A.1 and A.2 as well as Figures A.1 and A.2 show values of both e^x and $\ln(x)$ for various values of x.

The natural logarithmic transformation is the inverse transformation of the exponential. If $y = e^x$, then $\ln(y) = \ln(e^x)$ and $\ln(e^x) = x$. For example,

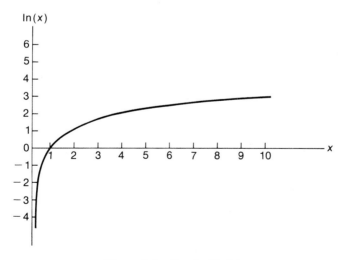

Figure A.1   Graph of ln (x)

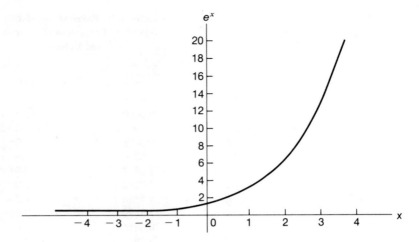

Figure A.2   Graph of e^x

if we let $y = e^x$ and let $x = 2.0$, then $y = e^{2.0} = 7.38906$. Now if we take the natural logarithm of y, $\ln(y)$, we see that $\ln(7.38906) = 2.0$ (as shown in Tables A.1 and A.2).

OPERATIONS WITH e AND THE NATURAL LOGARITHM

In this book we make use of three properties of the exponential and logarithmic transformations:

PROPERTY 1: $e^a \times e^b = e^{a+b}$. Example:

$$3^2 \times 3^3 = 9 \times 27 = 243$$

but

$$243 = 3^5 = 3^{2+3}$$

PROPERTY 2: $\ln(x \cdot y) = \ln(x) + \ln(y)$. Example:

$$\ln(2.71828 \times 1.64872) = \ln(2.71828) + \ln(1.64872)$$

From Table A.2 of natural logarithms we find that $1.0 + 0.5 = 1.5$. To verify that this value is correct, form the product of 2.71828 and 1.64872 (which equals 4.48169) and $\ln(4.48169) = 1.5$.

PROPERTY 3: $\ln(x/y) = \ln(x) - \ln(y)$. Example:

$$\ln\left(\frac{4.48169}{1.64872}\right) = \ln(4.48169) - \ln(1.64872)$$

$$= 1.5 - 0.5 = 1.0$$

Since $4.48169/1.64872 = 2.71828$, and $\ln(2.71828) = 1.0$, property 3 is verified.

One example of the use of the logarithmic transformation deals with the ratio of two quantities—for example, p and q. Suppose that the ratio of p and q is related to the quantities a, b, and c in the following multiplicative fashion:

$$\frac{p_{ij}}{q_{ij}} = c \times a_i \times b_j$$

Taking the natural logarithm of both sides of this equation yields

$$\ln\left(\frac{p_{ij}}{q_{ij}}\right) = \ln(c) + \ln(a_i) + \ln(b_j)$$

This equation is an additive relationship in terms of the logarithms of (p/q), c, a, and b. Hence taking the natural logarithm transforms a multiplicative relationship into an additive one, and techniques for analyzing additive relationships are well developed. Since procedures for analyzing multiplicative relationships are not developed to the same extent, we prefer to work with additive relationships whenever possible.

Setting $g_{ij} = \ln(p_{ij}/q_{ij})$, $d = \ln(c)$, $e_i = \ln(a_i)$, and $f_j = \ln(b_j)$ yields $g_{ij} = d + e_i + f_j$. Note that if e_i increases, this implies that g_{ij} also increases, which says that $\ln(p_{ij}/q_{ij})$ increases. This also means that p_{ij}/q_{ij} increases because, as can be seen from Figure A.1, $\ln(x)$ increases whenever x increases. Therefore an increase in the value of e_i means that the ratio of p to q also increases. We see the application of this idea throughout the book.

We also use both e and the natural logarithm with column vectors. If $\mathbf{F}$ is a column vector, for example, the natural logarithm of $\mathbf{F}$ is

$$\ln(\mathbf{F})_{(r \times 1)} = \begin{bmatrix} \ln(f_1) \\ \ln(f_2) \\ \vdots \\ \ln(f_r) \end{bmatrix}$$

SUMMARY

This appendix provides a brief introduction to matrix applications. We have discussed matrix arithmetic—especially matrix multiplication—and have applied matrices in the solution of linear equations. The appendix also introduces the exponential and logarithmic transformations, both of which are useful in the weighted least squares approach to categorical data.

These are the key topics covered in this appendix:

- *Matrix definitions:* A rectangular array of symbols.
- *Matrix arithmetic:*

$$\text{Matrix addition: } c_{ij} = a_{ij} + b_{ij}$$

$$\text{Matrix subtraction: } c_{ij} = a_{ij} - b_{ij}$$

$$\text{Matrix multiplication: } c_{ij} = \sum a_{ik} b_{kj}$$

- *Matrix inverse:* Analogous to division.
- *System of linear equations—scalar presentation:* Substitution method of solution.
- *Systems of linear equations—matrix presentation:* $x = A^{-1}b$; exactly equivalent to scalar solution.
- e^x *and* $\ln(x)$: Two transformations useful in the weighted-least-squares approach to categorical data; note that $\ln(e^x) = x$ and $e^{\ln(x)} = x$.
- *Operations:*

$$e^a \times e^b = e^{a+b}$$

$$\ln(x \times y) = \ln(x) + \ln(y)$$

$$\ln(x/y) = \ln(x) - \ln(y)$$

These are the key mathematical definitions and operations needed for this book. Appendix B provides the essential linear model material.

EXERCISES

A.1.

Let **X** be an 8 × 3 matrix and β be a 3 × 1 vector defined as

$$\underset{(8 \times 3)}{\mathbf{X}} = \begin{bmatrix} 1 & 1 & 1 \\ 1 & 1 & 1 \\ 1 & 1 & -1 \\ 1 & 1 & -1 \\ 1 & -1 & 1 \\ 1 & -1 & 1 \\ 1 & -1 & -1 \\ 1 & -1 & -1 \end{bmatrix} \quad \text{and} \quad \underset{(3 \times 1)}{\boldsymbol{\beta}} = \begin{bmatrix} \beta_0 \\ \beta_1 \\ \beta_2 \end{bmatrix}$$

a. Form $\mathbf{X'X}$ and $\mathbf{X}\boldsymbol{\beta}$.

b. Form $(\mathbf{X'X})^{-1}$.

c. Consider the equation $\mathbf{C}\boldsymbol{\beta} = \beta_1 - \beta_2$. What are the elements in $\mathbf{C}$ that satisfy this equation?

d. Given $\mathbf{X} = \mathbf{I}_8$ (the identity matrix), form $(\mathbf{X'X})$ and $(\mathbf{X'X})^{-1}$.

A.2.

Given

$$\mathbf{p}_{(4 \times 1)} = \begin{bmatrix} p_{11} \\ p_{12} \\ p_{21} \\ p_{22} \end{bmatrix}$$

consider the equation $p_{11}p_{22}/p_{12}p_{21} = 1$. Express this relationship in terms of the natural logarithm of p_{ij} (that is, $\ln(p_{ij})$).

A.3.

Using the vector $\mathbf{p}$ defined in Exercise A.2, determine what $\mathbf{A}$ in the equation $\mathbf{A}\ln(\mathbf{p}) = \mathbf{0}$ will produce the same equation derived in Exercise A.2.

A.4.

The expression

$$\mathbf{b} = (\mathbf{X'V}_F^{-1}\mathbf{X})^{-1}\mathbf{X'V}_F^{-1}\mathbf{F}$$

is the weighted least squares (WLS) solution for $\mathbf{b}$ given the linear model $\mathbf{F} = \mathbf{X}\boldsymbol{\beta}$. Let $\mathbf{X} = \mathbf{I}$, the identity matrix, and simplify the WLS expression.

Appendix B:
The Linear Model

Two widely used statistical techniques are the analysis of variance and linear regression. Both techniques are applications of the *linear model*, which is a method of expressing quantitatively the relationships between a dependent or response variable and a set of independent variables. When we use the linear model we are attempting to determine which, if any, of the independent variables are significantly related to the dependent variable and to discover the form of that relationship. We do not use linear model methodology because we believe that the world is linear. We use it because it is often a good approximation to the true relationships and provides a good point of departure for further study. Since the linear model is also an additive model that is tractable mathematically, its use is all the more attractive.

This appendix applies the linear model in the analysis of variance (ANOVA) setting. In particular we focus our discussion on the different types of coding or parameterizations possible to represent the model. Draper and Smith (1966) and Kleinbaum and Kupper (1978) are two sources for the application of the linear model to regression as well as to the analysis of variance.

The traditional presentation of the ANOVA procedure relies on a partitioning of the variance in the dependent variable and as a result it masks the model that is implied. Here we present both the usual ANOVA approach and the linear model approach and demonstrate their equivalence.

Note that we are considering only *balanced* data situations in this appendix—that is, there are the same number of observations for all combinations of levels of the independent variables. If sex is the independent variable, balance requires that there be the same number of females as males. If race and sex are the two independent variables in the study and black and white are the only racial categories of interest, balance requires that there be an equal number of black females, black males, white females, and white males.

TRADITIONAL APPROACH TO ANOVA

The data set we use in our development of the ANOVA procedure is artificial, but it is based on a real study. The question of interest is whether there is a difference in salary levels for female and male faculty members at a large university. The data for this investigation (Table B.1) result from a random sample of university faculty members. We begin with a consideration of the sex variable and incorporate the faculty rank variable later in the discussion. The following paragraphs present the rationale for the balanced one-way or one independent variable ANOVA procedure as well as a brief review of the calculations required to test the hypothesis of no difference between the mean female and male salaries. Hoel (1971) provides additional detail on ANOVA.

Table B.1   Monthly Salary, Sex, and Faculty Rank Data for 18 Faculty Members in the Same Department

FEMALES			MALES		
Faculty Number	*Monthly Salary ($)*	*Professorial Rank*	*Faculty Number*	*Monthly Salary ($)*	*Professorial Rank*
1	1700	Assistant	10	1400	Assistant
2	1700	Assistant	11	1600	Assistant
3	1600	Assistant	12	1300	Assistant
4	1800	Associate	13	1900	Associate
5	2100	Associate	14	2300	Associate
6	1900	Associate	15	2200	Associate
7	2500	Full	16	2900	Full
8	2600	Full	17	2800	Full
9	2600	Full	18	3000	Full

Let Y_i, $i = 1, 2, \ldots, 18$, represent the observed values for the salary variable with Y_1 to Y_9 being the observations for the female faculty and Y_{10} to Y_{18} representing the males' salaries. We assume that the Y_i are *independent* of one another, that they come from a *normal* distribution, and that the *variance* of the observations σ^2 is *constant* across gender.

In the ANOVA procedure we test the hypothesis that there is no difference between mean male and female salaries. The ANOVA procedure is based on the fact that the data provide two different ways of estimating the variance of the observations σ^2. The first method, the *between-group estimate*, is based on the relationship between the variance of an observation and the variance of a sample mean. This relationship is given by

$$\sigma_{\bar{y}}^2 = \frac{\sigma^2}{m} \tag{B.1}$$

where the variance of the mean of a sample of m observations is equal to the variance σ^2 divided by the sample size. Hence the sample means for the two sexes provide an estimate of $\sigma_{\bar{y}}^2$. Let $\bar{Y}_1$ and $\bar{Y}_2$ be the sample means for the female and male faculty members, respectively, and let $\bar{Y}$ be the mean for the entire sample. Then their sample variance is

$$S_{\bar{y}}^2 = \frac{\sum\limits_{i=1}^{2} (\bar{Y}_i - \bar{Y})^2}{2 - 1}$$

where $S_{\bar{y}}^2$ is the sample estimate of $\sigma_{\bar{y}}^2$. There are nine faculty members within each gender, and multiplication of $S_{\bar{y}}^2$ by 9 yields the between-group estimate of σ^2. This estimate is based on the assumption that males and females have the same population mean. If the sexes really have different means, then this estimate, $9S_{\bar{y}}^2$, is an overestimate of σ^2.

The second estimate, the *within-group estimate*, does not depend on whether the sexes have the same population mean. It examines the variation within each sex and then sums across sexes to obtain the estimate of σ^2. The sample variance for females is given by

$$S_1^2 = \frac{\sum\limits_{i=1}^{9} (Y_i - \bar{Y}_1)^2}{9 - 1}$$

and S_2^2 is defined similarly. The within-group estimate is given by

$$S^2 = \frac{S_1^2 + S_2^2}{2}$$

In summary, the between-group and within-group estimates of σ^2 are

Between: $$\frac{9 \cdot \sum\limits_{i=1}^{2} (\bar{Y}_i - \bar{Y})^2}{2} \tag{B.2}$$

Within:
$$\frac{\sum\limits_{i=1}^{2} S_i^2}{2} = \frac{\sum\limits_{i=1}^{9} (Y_i - \bar{Y}_1)^2 + \sum\limits_{i=10}^{18} (Y_i - \bar{Y}_2)^2}{8 \times 2} \tag{B.3}$$

The between-group estimate provides an estimate of σ^2 if the population means for each sex are the same. If the means are different, this between-group variance estimate overestimates σ^2. This fact can be used to test whether the sexes have equal means by forming the ratio of the between-group estimate to the within-group estimate. If this ratio is close to 1, it indicates that the sexes have similar means; if the ratio is large, it suggests that the means are different.

This ratio of two independent estimates of σ^2 follows the F distribution and thus we can use probability to define what is "large." The F distribution has two parameters, and they are the degrees of freedom associated with each variance estimate. For the between-group estimate, we calculate the variance of two means; thus there are $2 - 1$ degrees of freedom for this estimate. For the within-group estimate, we find the variance of nine observations; thus there are $9 - 1$ degrees of freedom for each sex. Therefore there are $2 \times 8 = 16$ degrees of freedom associated with the within-group estimate. The tabulated values of the F distribution can be written as

$$F_{n_1, n_2, 1 - \alpha}$$

where n_1 represents the degrees of freedom associated with the numerator of the F ratio, n_2 represents the denominator degrees of freedom, and α represents the probability level of interest.

We will test the hypothesis of no difference between the sex means at the 0.05 level—in other words, there is a 5 percent chance of falsely rejecting the hypothesis of no difference. Only a too large value of the F ratio will cause us to reject the hypothesis of no difference. The decision rule is:

> Reject the hypothesis of no difference if the calculated ratio is greater than or equal to ($\geq$) the tabulated F value, $F_{1, 16, 0.95} = 4.49$; fail to reject the hypothesis of no difference if the calculated ratio is less than ($<$) the $F_{1, 16, 0.95}$ value of 4.49.

We then calculate the F value:

$$\bar{Y}_1 = \sum_{i=1}^{9} \frac{Y_i}{9} = \frac{18,500}{9} = 2055.56$$

$$\bar{Y}_2 = \sum_{i=10}^{18} \frac{Y_i}{9} = \frac{19,400}{9} = 2155.56$$

$$\bar{Y} = \sum_{i=1}^{18} \frac{Y_i}{18} = \frac{37,900}{18} = 2105.56$$

$$S_{\bar{y}}^2 = \sum_{i=1}^{2} \frac{(\bar{Y}_i - \bar{Y})^2}{2} = 5000$$

$$9 \cdot S_{\bar{y}}^2 = 45,000$$

The between-group estimate of σ^2 is 45,000. The within-group estimate is

$$\frac{\sum_{i=1}^{9} (Y_i - \bar{Y}_1)^2 + \sum_{i=10}^{18} (Y_i - \bar{Y}_2)^2}{8 \times 2} = \frac{4,724,444.44}{16}$$

$$= 295,277.78$$

The between-group estimate of σ^2 is much smaller than the within-group estimate: 45,000 versus 295,277.78. The test statistic is the ratio of these two estimates:

$$F = \frac{45,000}{295,277.78} = 0.15$$

Since $0.15 < 4.49$ ($F_{1, 16, 0.95}$) we fail to reject the hypothesis of no difference in female and male mean salaries.

These results are usually summarized in an ANOVA table as follows:

Source of Variation	DF	Sum of Squares	Mean Square	F Ratio
Between	1	45,000.00	45,000	0.15
Within	16	4,724,444.44	295,277.78	
Total	17	4,769,444.44		

In this table the entries in the "mean square" column result from dividing the entries in the "sum of squares" column by their respective degrees of freedom. The F ratio is the ratio of the between to the within mean square. The creation of the ANOVA table and the decision whether or not to reject the hypothesis usually completes the standard presentation of ANOVA. In the next section we discuss the linear model approach to the ANOVA procedure.

LINEAR MODEL APPROACH TO ANOVA

This section focuses on the underlying model for the ANOVA procedure and shows how to test the hypothesis of no difference between the sexes and other hypotheses. The model implied by the ANOVA procedure is

Salary = constant + effect of gender on salary level

If we let

$$Y = \text{salary}$$
$$\beta_0 = \text{constant}$$
$$\beta_1 = \text{female effect}$$
$$\beta_2 = \text{male effect}$$

the model can now be written as

$$Y = \beta_0 + \text{either } \beta_1 \text{ or } \beta_2$$

To indicate which sex effect should be included in the equation, we define

$$X_{ij} = \begin{cases} 1 \text{ if the } i\text{th person is of sex } j & i = 1, 2, \ldots, 18 \\ 0 \text{ otherwise} & j = 1, 2 \end{cases}$$

and the model is

$$Y_i = \beta_0 + X_{i1}\beta_1 + X_{i2}\beta_2$$

We now examine the first, the ninth, and the seventeenth data points to ensure that the definition of the X_{ij}'s is clear. These data are:

i	Y_i	Sex	X_{i1} (Female)	X_{i2} (Male)
1	1700	female	1	0
9	2600	female	1	0
17	2800	male	0	1

where the X_{ij} simply indicate which β_j to include in the model. We can use the matrix notation from Appendix A to represent this situation succinctly, where the matrix equation expresses the n equations of the form

$$Y_i = \beta_0 + X_{i1}\beta_1 + X_{i2}\beta_2 \qquad i = 1, 2, \ldots, 18 \tag{B.4}$$

The matrix equation is

$$\underset{(18 \times 1)}{\mathbf{Y}} = \underset{(3 \times 1)}{\mathbf{X} \, \boldsymbol{\beta}} \tag{B.5}$$

It is easy to determine that $\mathbf{Y}$ is the vector containing the observed values of the dependent variable and $\boldsymbol{\beta}$ is the vector whose elements are the three parameters of the model:

$$\underset{(18 \times 1)}{\mathbf{Y}} = \begin{bmatrix} Y_1 \\ Y_2 \\ \vdots \\ Y_{18} \end{bmatrix} \qquad \underset{(3 \times 1)}{\boldsymbol{\beta}} = \begin{bmatrix} \beta_0 \\ \beta_1 \\ \beta_2 \end{bmatrix}$$

The definition of $\mathbf{X}$ is not quite so obvious, however. By examining (B.5) we see that $\mathbf{X}$ has 18 rows and 3 columns (it must have 3 columns to premultiply $\boldsymbol{\beta}$), and we know that the X_{ij}'s corresponding to β_1 and β_2 are either 1 or 0. The first column in $\mathbf{X}$ corresponds to β_0; the coefficient of β_0 in the equations is always 1. Therefore if we allow the first column of $\mathbf{X}$ to be all 1's, the $\mathbf{X}$ matrix is

$$\underset{(18 \times 3)}{\mathbf{X}} = \begin{bmatrix} 1 & 1 & 0 \\ 1 & 1 & 0 \\ 1 & 1 & 0 \\ 1 & 1 & 0 \\ 1 & 1 & 0 \\ 1 & 1 & 0 \\ 1 & 1 & 0 \\ 1 & 1 & 0 \\ 1 & 1 & 0 \\ 1 & 0 & 1 \\ 1 & 0 & 1 \\ 1 & 0 & 1 \\ 1 & 0 & 1 \\ 1 & 0 & 1 \\ 1 & 0 & 1 \\ 1 & 0 & 1 \\ 1 & 0 & 1 \\ 1 & 0 & 1 \end{bmatrix}$$

If we examine the first, the ninth, and the seventeenth rows of $Y = X\beta$, that is,

$$\begin{bmatrix} Y_1 \\ Y_9 \\ Y_{17} \end{bmatrix} = \begin{bmatrix} 1 & 1 & 0 \\ 1 & 1 & 0 \\ 1 & 0 & 1 \end{bmatrix} \begin{bmatrix} \beta_0 \\ \beta_1 \\ \beta_2 \end{bmatrix}$$

the multiplication of β by these three rows of X yields

$$Y_1 = \beta_0 + \beta_1$$
$$Y_9 = \beta_0 + \beta_1$$
$$Y_{17} = \beta_0 + \beta_2$$

which is what we wanted it to represent. Hence the definition of X to represent the desired model is not difficult in this situation.

This model is close to complete; however, we really do not expect all the faculty members of the same sex to have exactly the same salary. Although there may be some random variation within the same sex, we expect it to be small if the model is appropriate. Let e_i represent this random error term for the ith faculty member. The model is now

$$Y_i = \beta_0 + X_{i1}\beta_1 + X_{i2}\beta_2 + e_i \tag{B.6}$$

where we assume that the error terms are independent of one another and follow a normal distribution with a mean of zero and a constant variance. To represent (B.6) completely in matrix terms we must define the error vector:

$$\mathop{\mathbf{e}}_{(n \times 1)} = \begin{bmatrix} e_1 \\ e_2 \\ \vdots \\ e_n \end{bmatrix}$$

Inclusion of $\mathbf{e}$ with (B.5) yields

$$\mathop{\mathbf{Y}}_{(n \times 1)} = \mathop{\mathbf{X}}_{(n \times p)} \mathop{\boldsymbol{\beta}}_{(p \times 1)} + \mathop{\mathbf{e}}_{(n \times 1)} \tag{B.7}$$

where p is the number of parameters in β.

Now that we have defined the model, it is of interest to estimate the sex effects and to determine whether they are the same. The standard procedure for estimating $\boldsymbol{\beta}$ is to use *least squares* estimation, which is based on minimizing the sum of squares of the error term—that is, minimizing $\sum\limits_{i=1}^{n} e_i^2$. The estimate of $\boldsymbol{\beta}$ that minimizes $\sum\limits_{i=1}^{n} e_i^2$ is given by

$$\mathbf{b} = (\mathbf{X'X})^{-1}\mathbf{X'Y} \tag{B.8}$$

We now apply this equation to our example. The first step is to form the indicated products:

$$\underset{(3 \times 18)(18 \times 1)}{\mathbf{X'} \quad \mathbf{Y}} = \begin{bmatrix} 1 & 1 & 1 & 1 & 1 & 1 & 1 & 1 & 1 & 1 & 1 & 1 & 1 & 1 & 1 & 1 & 1 & 1 \\ 1 & 1 & 1 & 1 & 1 & 1 & 1 & 1 & 1 & 0 & 0 & 0 & 0 & 0 & 0 & 0 & 0 & 0 \\ 0 & 0 & 0 & 0 & 0 & 0 & 0 & 0 & 0 & 1 & 1 & 1 & 1 & 1 & 1 & 1 & 1 & 1 \end{bmatrix} \begin{bmatrix} 1700 \\ 1700 \\ 1600 \\ 1800 \\ 2100 \\ 1900 \\ 2500 \\ 2600 \\ 2600 \\ 1400 \\ 1600 \\ 1300 \\ 1900 \\ 2300 \\ 2200 \\ 2900 \\ 2800 \\ 3000 \end{bmatrix}$$

$$\underset{(3 \times 1)}{\mathbf{X'Y}} = \begin{bmatrix} 37,900 \\ 18,500 \\ 19,400 \end{bmatrix}$$

$$\underset{(3 \times 18)(18 \times 3)}{\mathbf{X'} \quad \mathbf{X}} = \begin{bmatrix} 18 & 9 & 9 \\ 9 & 9 & 0 \\ 9 & 0 & 9 \end{bmatrix}$$

The next step is to find $(\mathbf{X'X})^{-1}$; before doing so, however, we examine $\mathbf{X'X}$ to see if it is singular.

In many cases it is possible to identify a singular $\mathbf{X'X}$ matrix by examining the $\mathbf{X}$ matrix itself. In this case the last two columns of $\mathbf{X}$ sum to the first column; a linear relationship exists among the columns of $\mathbf{X}$ and causes $\mathbf{X'X}$ to be singular. When this occurs, it means that we cannot uniquely determine estimates of all the parameters in the model because $\mathbf{X'X}$ cannot be inverted. In the next section we show a solution to this situation.

CODING METHODS

We are interested in whether there are significant differences between the two sex effects and not in the values of the sex effects. With this in mind, we discuss three assumptions, or reparameterizations, each of which will allow us to continue with the linear model approach to ANOVA:

1. Set the constant term β_0 equal to zero. Thus the effects are measured from zero and not from the constant term.
2. Set one of the sex effects equal to zero and measure the other effects from it.
3. Measure the sex effects from their mean. This means that the sum of their effects is equal to zero.

The choice of assumption has no impact on the tests of hypotheses, and hence it is arbitrary in that sense. Now let us see how each of these assumptions affects the $\mathbf{X}$ matrix and the $\boldsymbol{\beta}$ vector. We use subscripts on the $\mathbf{X}$ matrix and $\boldsymbol{\beta}$ vector to distinguish the different models.

METHOD 1. The β_0 term is set equal to zero or, equivalently, it is deleted from the model. The model becomes

$$Y_i = X_{i1}\beta_{11} + X_{i2}\beta_{12} + e_i$$

In this case we need only delete the first column of $\mathbf{X}$ corresponding to the deletion of β_0. This new model is

$$\underset{(18 \times 1)}{\mathbf{Y}} = \underset{(18 \times 2)(2 \times 1)}{\mathbf{X}_1 \, \boldsymbol{\beta}_1} + \underset{(18 \times 1)}{\mathbf{e}}$$

with

$$\underset{(2 \times 1)}{\boldsymbol{\beta}_1} = \begin{bmatrix} \beta_{11} \\ \beta_{12} \end{bmatrix} = \begin{bmatrix} \text{effect of females} \\ \text{effect of males} \end{bmatrix}$$

and

$$\mathbf{X}_1 \atop (18 \times 2) =
\begin{bmatrix}
1 & 0 \\
1 & 0 \\
1 & 0 \\
1 & 0 \\
1 & 0 \\
1 & 0 \\
1 & 0 \\
1 & 0 \\
1 & 0 \\
0 & 1 \\
0 & 1 \\
0 & 1 \\
0 & 1 \\
0 & 1 \\
0 & 1 \\
0 & 1 \\
0 & 1 \\
0 & 1
\end{bmatrix}$$

$$\mathbf{X}_1'\mathbf{Y} = \begin{bmatrix} 18{,}500 \\ 19{,}400 \end{bmatrix} \qquad \mathbf{X}_1'\mathbf{X}_1 = \begin{bmatrix} 9 & 0 \\ 0 & 9 \end{bmatrix}$$

$$(\mathbf{X}_1'\mathbf{X}_1)^{-1} = \begin{bmatrix} \frac{1}{9} & 0 \\ 0 & \frac{1}{9} \end{bmatrix}$$

$$\mathbf{b}_1 \atop (2 \times 1) = (\mathbf{X}_1'\mathbf{X}_1)^{-1}\mathbf{X}_1'\mathbf{Y}$$

$$\begin{bmatrix} b_{11} \\ b_{12} \end{bmatrix} = \begin{bmatrix} \frac{1}{9} & 0 \\ 0 & \frac{1}{9} \end{bmatrix} \begin{bmatrix} 18{,}500 \\ 19{,}400 \end{bmatrix} = \begin{bmatrix} 2055.56 \\ 2155.56 \end{bmatrix} \tag{B.9}$$

Note that these values are the average salaries for females and males. Hence we have successfully removed the linear dependency that existed previously and have obtained estimates of the two sex effects.

METHOD 2. Set one of the effects equal to zero and measure the other effects from it. Upon setting β_2 equal to zero, the new model is

$$Y_i = \beta_{20} + X_{i1}\beta_{21} + e_i$$

In this case we need only delete the last column of $\mathbf{X}$ to correspond to setting β_2 equal to zero. The model is

$$\mathbf{Y} = \mathbf{X}_2\boldsymbol{\beta}_2 + \mathbf{e}$$

with

$$\boldsymbol{\beta}_2 = \begin{bmatrix} \beta_{20} \\ \beta_{21} \end{bmatrix} = \begin{bmatrix} \text{constant term} \\ \text{dummy effect of females} \end{bmatrix}$$

We use the term *dummy* because the female effect is measured relative to the males.

$$\mathbf{X}_2 = \begin{bmatrix} 1 & 1 \\ 1 & 1 \\ 1 & 1 \\ 1 & 1 \\ 1 & 1 \\ 1 & 1 \\ 1 & 1 \\ 1 & 1 \\ 1 & 1 \\ 1 & 0 \\ 1 & 0 \\ 1 & 0 \\ 1 & 0 \\ 1 & 0 \\ 1 & 0 \\ 1 & 0 \\ 1 & 0 \\ 1 & 0 \end{bmatrix}$$

$$\mathbf{X}_2'\mathbf{Y} = \begin{bmatrix} 37{,}900 \\ 18{,}500 \end{bmatrix} \qquad \mathbf{X}_2'\mathbf{X}_2 = \begin{bmatrix} 18 & 9 \\ 9 & 9 \end{bmatrix}$$

$$(\mathbf{X}_2'\mathbf{X}_2)^{-1} = \begin{bmatrix} \frac{1}{9} & -\frac{1}{9} \\ -\frac{1}{9} & \frac{2}{9} \end{bmatrix}$$

$$\mathbf{b}_2 = (\mathbf{X}_2'\mathbf{X}_2)^{-1}\mathbf{X}_2'\mathbf{Y}$$

$$= \begin{bmatrix} \frac{1}{9} & -\frac{1}{9} \\ -\frac{1}{9} & \frac{2}{9} \end{bmatrix} \begin{bmatrix} 37{,}900 \\ 18{,}500 \end{bmatrix} = \begin{bmatrix} 2155.56 \\ -100.00 \end{bmatrix}$$

$$\begin{bmatrix} \mathbf{b}_{20} \\ \mathbf{b}_{21} \end{bmatrix} = \begin{bmatrix} 2155.56 \\ -100.00 \end{bmatrix} = \begin{bmatrix} \mathbf{b}_{12} \\ \mathbf{b}_{11} - \mathbf{b}_{12} \end{bmatrix} \qquad (\text{B.10})$$

Here the constant term is now the mean salary for males and the dummy female effect is the difference between the female and male mean salaries.

METHOD 3.   Measure the effects from their mean. This implies that the sum of the effects is equal to zero because the sum of deviations about the mean is zero. Given

$$\beta_{31} \quad = \quad \beta_{11} - \frac{\beta_{11} + \beta_{12}}{2}$$

$$\beta_{32} \quad = \quad \beta_{12} - \frac{\beta_{11} + \beta_{12}}{2}$$

then the sum is

$$\beta_{31} + \beta_{32} = \beta_{11} + \beta_{12} - 2(\beta_{11} + \beta_{12})/2 = 0$$

This means that $\beta_{32} = -\beta_{31}$ and therefore we require only β_{30} and β_{31} in the model because we can formulate β_{32} in terms of β_{31}. Hence the new model is

$$Y_i = \beta_{30} + X_{i1}\beta_{31} + e_i$$

The model differs from the previous model because the X_{ij} are no longer just 1's and 0's. We have to allow the X_{ij}'s to have the value of -1 as well, because we represent the effect of males by $-\beta_{31}$. Therefore the model is

$$\mathbf{Y} = \mathbf{X}_3\boldsymbol{\beta}_3 + \mathbf{e}$$

where $\boldsymbol{\beta}_3$ and $\mathbf{X}_3$ are

$$\boldsymbol{\beta}_3 = \begin{bmatrix} \beta_{30} \\ \beta_{31} \end{bmatrix} = \begin{bmatrix} \text{constant term} \\ \text{differential effect of females} \end{bmatrix}$$

We use the term *differential effect* here because the effect is measured relative to the mean of the sex effects.

$$X_3 = \begin{bmatrix} 1 & 1 \\ 1 & 1 \\ 1 & 1 \\ 1 & 1 \\ 1 & 1 \\ 1 & 1 \\ 1 & 1 \\ 1 & 1 \\ 1 & 1 \\ 1 & -1 \\ 1 & -1 \\ 1 & -1 \\ 1 & -1 \\ 1 & -1 \\ 1 & -1 \\ 1 & -1 \\ 1 & -1 \\ 1 & -1 \end{bmatrix}$$

The last nine rows of X_3 multiplied by β_3 yield $\beta_{30} - \beta_{31} = \beta_{30} + \beta_{32}$. Now

$$X_3'Y = \begin{bmatrix} 37{,}900 \\ -900 \end{bmatrix} \qquad X_3'X_3 = \begin{bmatrix} 18 & 0 \\ 0 & 18 \end{bmatrix}$$

$$(X_3'X_3)^{-1} = \begin{bmatrix} \frac{1}{18} & 0 \\ 0 & \frac{1}{18} \end{bmatrix}$$

$$b_3 = (X_3'X_3)^{-1} X_3'Y$$

$$= \begin{bmatrix} \frac{1}{18} & 0 \\ 0 & \frac{1}{18} \end{bmatrix} \begin{bmatrix} 37{,}900 \\ -900 \end{bmatrix} = \begin{bmatrix} 2105.56 \\ -50.00 \end{bmatrix}$$

$$\begin{bmatrix} b_{30} \\ b_{31} \end{bmatrix} = \begin{bmatrix} 2105.56 \\ -50.00 \end{bmatrix} = \begin{bmatrix} (b_{11} + b_{12})/2 \\ b_{11} - (b_{11} + b_{12})/2 \end{bmatrix} \qquad \text{(B.11)}$$

The constant term is the average of the mean female and male salaries, and the differential female effect is the female mean minus the average of the female and male means. All three types of coding are used in the book. Select the coding or reparameterization that you prefer.

We will be working with the last model for the remaining analyses. The estimated effects tell us that females tend to have a salary which is lower than average by \$50.00 ($\mathbf{b}_{31} = -50.00$). The effect of males is found by taking minus the female effect, which yields plus \$50.00. Hence it appears that males have a slightly higher than average salary in this artifical data set. The next step in the analysis is to test whether the effects differ significantly.

TESTING HYPOTHESES

Before we discuss the testing of hypotheses, let us reiterate the assumptions about the dependent variable that we make for ANOVA procedure. We assume that the observed values of the salary variable are independent of one another, that they follow a normal distribution, and that the female and male salaries have the same variance. If these assumptions are satisfied, we may test the hypothesis.

Tests of hypotheses are expressed as linear combinations of the β_i in the model. A way of representing linear combinations of the β_i is to premultiply $\boldsymbol{\beta}$ by a matrix. The elements of the matrix are the coefficients of the β_i's in the hypothesis being tested. If we let $\mathbf{C}$ be the matrix of coefficients of the β_i, the hypothesis under study can be written as

$$H_0: \quad \mathbf{C}\boldsymbol{\beta} = \mathbf{0}$$

The $\mathbf{C}$ matrix gives great flexibility in the choice of hypothesis to be tested.

With this introduction to the test of hypotheses, let us consider the salary data. We will use the model based on the third method: $\mathbf{X}_3\boldsymbol{\beta}_3$. Because we have settled on a model, we will drop the subscript from the $\mathbf{X}$ and $\boldsymbol{\beta}$. The model is then

$$Y_i = \beta_0 + X_{i1}\beta_1 + e_i$$

where β_1 is the female effect measured from the mean of the female and male effects. The hypothesis of interest is that of no difference in the sex effects. Since in this reparameterization $\beta_2 = -\beta_1$, the only way that β_1 and β_2 can be equal is for them both to be zero. Therefore the hypothesis of no sex effect is expressed by

$$H_0: \quad \beta_1 = 0 \tag{B.12}$$

It is not necessary to test a hypothesis about β_2; if β_1 is zero, then β_2 is also zero.

In this hypothesis, β_0 is not shown—that is, it has a coefficient of zero—and β_1 has a coefficient of 1. Therefore the $\mathbf{C}$ matrix that expresses this same hypothesis in $H_0: \quad \mathbf{C}\boldsymbol{\beta} = \mathbf{0}$ is

$$\mathbf{C} = \begin{bmatrix} 0 & 1 \end{bmatrix}$$

because $C\beta$ is

$$[0 \quad 1]\begin{bmatrix} \beta_0 \\ \beta_1 \end{bmatrix} = [\beta_1]$$

which is the left side of the hypothesis shown in (B.12).

Recall that this is the same hypothesis we tested before. We again use the F distribution to test this hypothesis. As we have seen, the F statistic has two parameters, and they are the degrees of freedom associated with the numerator and denominator of the F ratio respectively. The numerator degrees of freedom is equal to the number of rows in the C matrix (in this case 1); the denominator degrees of freedom is equal to the number of observations minus the number of rows in β (in this case 16).

The matrix form of the F ratio is

$$F = \frac{b'C'[C(X'X)^{-1}C']^{-1}Cb/c}{(Y'Y - b'X'Y)/(n - p)}$$

where c is the number of rows in the C matrix and p is the number of columns in β. We perform these calculations by hand here to demonstrate what the computer does. We have found that

$$b = \begin{bmatrix} 2105.56 \\ -50.00 \end{bmatrix}$$

Therefore

$$Cb = [0 \quad 1]\begin{bmatrix} 2105.56 \\ -50.00 \end{bmatrix} = [-50.00]$$

$$(X'X)^{-1} = \begin{bmatrix} \frac{1}{18} & 0 \\ 0 & \frac{1}{18} \end{bmatrix}$$

$$C(X'X)^{-1} = [0 \quad 1]\begin{bmatrix} \frac{1}{18} & 0 \\ 0 & \frac{1}{18} \end{bmatrix} = [0 \quad \frac{1}{18}]$$

$$C(X'X)^{-1}C' = [0 \quad \frac{1}{18}]\begin{bmatrix} 0 \\ 1 \end{bmatrix} = [\frac{1}{18}]$$

$$[C(X'X)^{-1}C']^{-1} = [18]$$

$$bC'[C(X'X)^{-1}C']^{-1}Cb = [-50.00][18][-50.00] = [45,000]$$

$$Y'Y = [84,570,000]$$

$$b'X'Y = [2105.56 \quad -50.00]\begin{bmatrix} 37,900 \\ -900 \end{bmatrix}$$

$$= [79,845,555.56]$$

$$\mathbf{Y'Y} - \mathbf{b'X'Y} = [84{,}570{,}000 - 79{,}845{,}555.56]$$
$$= [4{,}724{,}444.44]$$

$$F = \frac{45{,}000/1}{4{,}724{,}444.44/16} = 0.15$$

These calculations produce the same value of the F statistic that we arrived at before. Based on this F ratio of 0.15 being less than the ninety-fifth percentile of the F distribution ($F_{1,16,0.95} = 4.49$), we fail to reject the hypothesis of no sex effect. There is no significant difference in the mean salaries by sex.

There are other hypotheses that we could test, but we will forgo them in order to introduce the faculty rank variable into the model. Since we wanted to start with a simple model, we initially considered only the sex variable. Now we are ready for the model with two independent variables or factors.

TWO-WAY ANOVA

The model becomes

$$\text{Salary} = \text{constant} + \text{sex effect}$$
$$+ \text{faculty rank effect} + \text{random error}$$

or

$$Y_i = \beta_0 + \text{female effect or male effect}$$
$$+ \text{effect of assistant professor or effect of associate}$$
$$\text{professor or effect of full professor} + e_i$$

We use the same method of coding as before—that is, measuring the effects of the independent variables from their mean value. We have the following definitions:

Y = salary

β_0 = constant term

β_1 = differential effect of females

β_2 = differential effect of assistant professors

β_3 = differential effect of associate professors

e = random error term

The differential effect of males is

$$-\beta_1$$

and the differential effect of full professors is

$$-\beta_2 - \beta_3$$

The model is

$$Y_i = \beta_0 + X_{i1}\beta_1 + X_{i2}\beta_2 + X_{i3}\beta_3 + e_i$$

The **X** matrix has 18 rows and 4 columns:

$$\mathbf{X} = \begin{bmatrix} 1 & 1 & 1 & 0 \\ 1 & 1 & 1 & 0 \\ 1 & 1 & 1 & 0 \\ 1 & 1 & 0 & 1 \\ 1 & 1 & 0 & 1 \\ 1 & 1 & 0 & 1 \\ 1 & 1 & -1 & -1 \\ 1 & 1 & -1 & -1 \\ 1 & 1 & -1 & -1 \\ 1 & -1 & 1 & 0 \\ 1 & -1 & 1 & 0 \\ 1 & -1 & 1 & 0 \\ 1 & -1 & 0 & 1 \\ 1 & -1 & 0 & 1 \\ 1 & -1 & 0 & 1 \\ 1 & -1 & -1 & -1 \\ 1 & -1 & -1 & -1 \\ 1 & -1 & -1 & -1 \end{bmatrix}$$

The first three faculty members are female assistant professors and the first three rows of **Xβ** yield

$$1 \cdot \beta_0 + 1 \cdot \beta_1 + 1 \cdot \beta_2 + 0 \cdot \beta_3$$

that is, the constant plus the differential female effect plus the differential assistant professor effect. The next three rows of **Xβ** yield

$$1 \cdot \beta_0 + 1 \cdot \beta_1 + 0 \cdot \beta_2 + 1 \cdot \beta_3$$

that is, the constant plus the differential female effect plus the differential associate professor effect. The next three rows of **Xβ**, rows 7 to 9, yield

$$1 \cdot \beta_0 + 1 \cdot \beta_1 - 1 \cdot \beta_2 - 1 \cdot \beta_3$$

that is, the constant plus the differential female effect minus the differential assistant professor effect minus the differential associate professor effect. Recalling that the differential full professor effect equals minus the differential assistant professor effect minus the differential associate professor effect, we see that this $\mathbf{X}$ matrix is producing the correct combinations of the β_i's.

We wish to find the $\mathbf{b}$ that characterizes the relationship between $\mathbf{Y}$ and the factors. Recall that $\mathbf{b} = (\mathbf{X'X})^{-1}\mathbf{X'Y}$; therefore we must find $(\mathbf{X'X})$, $(\mathbf{X'X})^{-1}$, and $\mathbf{X'Y}$. The computer could calculate these values for us, but we demonstrate the calculations one last time:

$$(\mathbf{X'X}) = \begin{bmatrix} 18 & 0 & 0 & 0 \\ 0 & 18 & 0 & 0 \\ 0 & 0 & 12 & 6 \\ 0 & 0 & 6 & 12 \end{bmatrix}$$

$$\mathbf{X'Y} = \begin{bmatrix} 37{,}900 \\ -900 \\ -7100 \\ -4200 \end{bmatrix}$$

$$(\mathbf{X'X})^{-1} = \begin{bmatrix} \frac{1}{18} & 0 & 0 & 0 \\ 0 & \frac{1}{18} & 0 & 0 \\ 0 & 0 & \frac{1}{9} & -\frac{1}{18} \\ 0 & 0 & -\frac{1}{18} & \frac{1}{9} \end{bmatrix}$$

$$\mathbf{b} = (\mathbf{X'X})^{-1}\mathbf{X'Y}$$

$$= \begin{bmatrix} 2105.56 \\ -50.00 \\ -555.56 \\ -72.22 \end{bmatrix}$$

Note that the constant and the sex effects have not changed. Females still have a salary slightly less than males while assistant professors have a salary more than \$500 below the average and associate professors are slightly below the average. The effect of being a full professor is

$$-b_2 - b_3 = -(-555.56) - (-72.22) = 627.78$$

Although this examination of the coefficients suggests that there are differences among the salaries by faculty rank, it is necessary to test whether these differences are statistically significant or not. To test for the existence of a faculty rank effect, the hypothesis is expressed by

$$H_0: \quad \beta_2 = 0$$
$$\beta_3 = 0$$

That is, if both β_2 and β_3 are zero, the effect of being a full professor must also be zero. The coefficients of β_0, β_1, and β_3 are zero in the first equation and the coefficients of β_0, β_1, and β_2 are zero in the second equation. Therefore the **C** matrix necessary to produce this hypothesis is

$$\mathbf{C} = \begin{bmatrix} 0 & 0 & 1 & 0 \\ 0 & 0 & 0 & 1 \end{bmatrix}$$

If we perform this test with a probability of falsely rejecting H_0 equal to 0.05, the critical value from the F tables is $F_{2,16,0.95} = 3.63$. If the calculated F ratio is greater than or equal to 3.63, we reject H_0; otherwise we fail to reject H_0. The F ratio for this hypothesis is 62.38; therefore we reject the hypothesis of no mean salary difference among the three faculty ranks.

This concludes our discussion of hypothesis testing. We could test additional hypotheses, but it is not our purpose to portray the actual analyses one would perform with this set of data. Rather, we wish to demonstrate the different types of coding, the testing of hypotheses, and the concept of interaction. We therefore turn now to an example of interaction without further discussion of the results of this model.

INTERACTION

We say that there is *interaction* between the sex and faculty rank variables if the effects of the faculty ranks are different for females and males. Figure B.1a–d shows some examples of interaction.

In Figure B.1a we see that the lines showing the relationship between salary and faculty rank for each sex are parallel. This means that the effect of faculty rank is the same for each sex and hence no interaction is present. In Figure B.1b we see that salary decreases as the rank increases for females whereas salary increases with faculty rank for males. Since the effect of faculty rank varies by sex, we say that there is interaction between faculty rank and sex. In fact, Figure B.1b shows that it makes no sense to talk about a faculty rank effect or a sex effect because it depends on the level of the two independent variables taken together. In Figures B.1c and d we also see that the effect of faculty rank varies by sex. The lines do not cross in the same way as in Figure B.1b, however, so the interaction here is of *degree* rather than a crossover type. When there is an interaction of degree (Figures B.1c and d), we can still talk

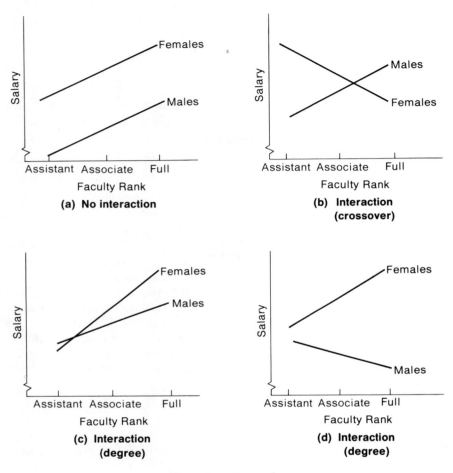

Figure B.1   Interaction

about the effects of faculty rank or sex. In Figure B.1*c* faculty rank varies positively with salary for both sexes and full professors have the highest salaries. In Figure B.1*d* there is a sex effect but not a faculty rank effect, because the effect of rank on salary depends on the individual's sex. These are the only logically possible patterns, and only one of them can exist for our data set. We now examine the data further to determine which of these patterns holds true.

There are six combinations of faculty rank by sex, each of which contributes an interaction term; hence we can have six additional β_i's to show the inclusion of interaction in the model. The β vector now contains 10 terms:

$$
\underset{(10 \times 1)}{\boldsymbol{\beta}} =
\begin{bmatrix}
\beta_0 \\
\beta_1 \\
\beta_2 \\
\beta_3 \\
\beta_4 \\
\beta_5 \\
\beta_6 \\
\beta_7 \\
\beta_8 \\
\beta_9
\end{bmatrix}
=
\begin{bmatrix}
\text{constant} \\
\text{differential effect of females} \\
\text{differential effect of assistant professors} \\
\text{differential effect of associate professors} \\
\text{interaction of females and assistant professors} \\
\text{interaction of females and associate professors} \\
\text{interaction of females and full professors} \\
\text{interaction of males and assistant professors} \\
\text{interaction of males and associate professors} \\
\text{interaction of males and full professors}
\end{bmatrix}
$$

Note that the last six columns of X select the appropriate interaction terms while the first four columns still select the constant and main effect terms just as before. For the first three faculty members, female assistant professors, the new columns in X multiplied by β now form

$$
1 \cdot \beta_4 + 0 \cdot \beta_5 + 0 \cdot \beta_6 + 0 \cdot \beta_7 + 0 \cdot \beta_8 + 0 \cdot \beta_9
$$

which represents the interaction of female and assistant professor effects. The corresponding X matrix is

$$
X =
\begin{bmatrix}
1 & 1 & 1 & 0 & 1 & 0 & 0 & 0 & 0 & 0 \\
1 & 1 & 1 & 0 & 1 & 0 & 0 & 0 & 0 & 0 \\
1 & 1 & 1 & 0 & 1 & 0 & 0 & 0 & 0 & 0 \\
1 & 1 & 0 & 1 & 0 & 1 & 0 & 0 & 0 & 0 \\
1 & 1 & 0 & 1 & 0 & 1 & 0 & 0 & 0 & 0 \\
1 & 1 & 0 & 1 & 0 & 1 & 0 & 0 & 0 & 0 \\
1 & 1 & -1 & -1 & 0 & 0 & 1 & 0 & 0 & 0 \\
1 & 1 & -1 & -1 & 0 & 0 & 1 & 0 & 0 & 0 \\
1 & 1 & -1 & -1 & 0 & 0 & 1 & 0 & 0 & 0 \\
1 & -1 & 1 & 0 & 0 & 0 & 0 & 1 & 0 & 0 \\
1 & -1 & 1 & 0 & 0 & 0 & 0 & 1 & 0 & 0 \\
1 & -1 & 1 & 0 & 0 & 0 & 0 & 1 & 0 & 0 \\
1 & -1 & 0 & 1 & 0 & 0 & 0 & 0 & 1 & 0 \\
1 & -1 & 0 & 1 & 0 & 0 & 0 & 0 & 1 & 0 \\
1 & -1 & 0 & 1 & 0 & 0 & 0 & 0 & 1 & 0 \\
1 & -1 & -1 & -1 & 0 & 0 & 0 & 0 & 0 & 1 \\
1 & -1 & -1 & -1 & 0 & 0 & 0 & 0 & 0 & 1 \\
1 & -1 & -1 & -1 & 0 & 0 & 0 & 0 & 0 & 1
\end{bmatrix}
\qquad (B.13)
$$

Examination of this matrix reveals the presence of singularities. For example, the first column is the sum of the last six columns. Moreover, the second column is the sum of columns 5, 6, and 7 minus the sum of columns 8, 9, and 10. From these two examples we can see that some of the columns are linear combinations of other columns in $\mathbf{X}$. This means that $\mathbf{X}$ is a singular matrix and we cannot find the inverse of $\mathbf{X}'\mathbf{X}$. To eliminate these problems, we will make assumptions similar to those made for both the sex and faculty rank variables. We now measure the three female by faculty rank effects from their mean and likewise measure the male and female assistant professor effects from their mean. These conditions imply that

$$\beta_4 + \beta_5 + \beta_6 = 0 \tag{B.14}$$

$$\beta_4 + \beta_7 \qquad = 0 \tag{B.15}$$

We make similar assumptions for males; that is,

$$\beta_7 + \beta_8 + \beta_9 = 0 \tag{B.16}$$

as well as for the other faculty ranks:

$$\beta_5 + \beta_8 \qquad = 0 \tag{B.17}$$

$$\beta_6 + \beta_9 \qquad = 0 \tag{B.18}$$

In summary the relationships are

$$\beta_6 = -\beta_4 - \beta_5 \qquad \text{(from B.14)}$$

or the differential female by full professor interaction effect equals minus the sum of the differential female by assistant professor and female by associate professor effects;

$$\beta_7 = -\beta_4 \qquad \text{(from B.15)}$$

or the differential male by assistant professor effect equals minus the differential female by assistant professor effect;

$$\beta_8 = -\beta_5 \qquad \text{(from B.17)}$$

or the differential male by associate professor effect equals minus the differential female by associate professor effect; and

$$\beta_9 = -\beta_7 - \beta_8 = -(-\beta_4) - (-\beta_5) \qquad \text{(from B.16)}$$
$$= \beta_4 + \beta_5$$

or the differential male by full professor effect equals the sum of the differential female by assistant professor and female by associate professor effects. Hence we are able to express β_6, β_7, β_8, and β_9 in terms of β_4 and β_5.

We now incorporate these relationships into our model and, as a result, there will be only six β_i's in the new model. The use of these relationships removes the sources of singularity in the $\mathbf{X}$ matrix. The choice of the assumptions is arbitrary, since we could have chosen other assumptions to remove the problems. However, the relationships used here are consistent with coding method 3. The new $\mathbf{X}$ and β are

$$\mathbf{X}\beta = \begin{bmatrix} 1 & 1 & 1 & 0 & 1 & 0 \\ 1 & 1 & 1 & 0 & 1 & 0 \\ 1 & 1 & 1 & 0 & 1 & 0 \\ 1 & 1 & 0 & 1 & 0 & 1 \\ 1 & 1 & 0 & 1 & 0 & 1 \\ 1 & 1 & 0 & 1 & 0 & 1 \\ 1 & 1 & -1 & -1 & -1 & -1 \\ 1 & 1 & -1 & -1 & -1 & -1 \\ 1 & 1 & -1 & -1 & -1 & -1 \\ 1 & -1 & 1 & 0 & -1 & 0 \\ 1 & -1 & 1 & 0 & -1 & 0 \\ 1 & -1 & 1 & 0 & -1 & 0 \\ 1 & -1 & 0 & 1 & 0 & -1 \\ 1 & -1 & 0 & 1 & 0 & -1 \\ 1 & -1 & 0 & 1 & 0 & -1 \\ 1 & -1 & -1 & -1 & 1 & 1 \\ 1 & -1 & -1 & -1 & 1 & 1 \\ 1 & -1 & -1 & -1 & 1 & 1 \end{bmatrix} \begin{bmatrix} \beta_0 \\ \beta_1 \\ \beta_2 \\ \beta_3 \\ \beta_4 \\ \beta_5 \end{bmatrix} \tag{B.19}$$

The elements in β represent the following terms:

$$\begin{bmatrix} \beta_0 \\ \beta_1 \\ \beta_2 \\ \beta_3 \\ \beta_4 \\ \beta_5 \end{bmatrix} = \begin{bmatrix} \text{constant} \\ \text{differential female effect} \\ \text{differential assistant professor effect} \\ \text{differential associate professor effect} \\ \text{differential female by assistant professor effect} \\ \text{differential female by associate professor effect} \end{bmatrix}$$

We now examine the seventh and seventeenth rows of $\mathbf{X}\beta$ to see if it provides the proper combination of the parameters. The seventh observation is a female who is a full professor. If we ignore the error terms, Y_7 would be represented by

Constant + differential female effect

+ differential full professor effect

+ differential interaction effect of female by full professor

Model $\mathbf{X}\boldsymbol{\beta}$ indicates that Y_7 is

$$\beta_0 + \beta_1 - \beta_2 - \beta_3 - \beta_4 - \beta_5$$

where $-\beta_2 - \beta_3$ represents the differential effect of a full professor. Since we have assumed that the differential female by faculty rank interaction parameters sum to zero, we know that the parameter representing female by full professor interaction can be expressed as minus the sum of the other female by rank interaction parameters. Therefore $-\beta_4 - \beta_5$ represents the differential female by full professor interaction effect. This $\mathbf{X}\boldsymbol{\beta}$ provides the proper combination of the parameters to represent Y_7.

In the same way, if we ignore the error term, Y_{17} can be expressed as

Constant + differential male effect

+ differential full professor effect

+ differential interaction effect of male by full professor

Model $\mathbf{X}\boldsymbol{\beta}$ indicates that Y_{17} is

$$\beta_0 - \beta_1 - \beta_2 - \beta_3 + \beta_4 + \beta_5$$

where $-\beta_1$ represents the differential effect of males and $(-\beta_2 - \beta_3)$ represents the differential full professor effect. Since we assumed that the sum of the differential male by faculty rank interaction effects is zero, we know that the male by full professor differential interaction effect can be expressed as minus the sum of the differential male by assistant professor and male by associate professor parameters. Therefore $(+\beta_4 + \beta_5)$ represents the differential male by full professor interaction effect; and $\mathbf{X}\boldsymbol{\beta}$ provides the proper coding for Y_{17} too. These examples demonstrate that the reparameterized model retains the basic structure of the model, but it has removed the sources of singularity present in the original formulation of $\mathbf{X}$ in Equation (B.13).

The reparameterization may appear complex, but there are relatively easy steps you can follow to form these differential interaction terms. If we examine the $\mathbf{X}$ matrix in Equation (B.19) carefully, we see that the first interaction column, the differential female by assistant professor effect (column 5 in $\mathbf{X}$), is the product of the differential female column and the differential assistant professor column (columns 2 and 3 in **X**). The other interaction columns in **X** can be produced in the same fashion—that is, by multiplying the columns in **X** corresponding to the appropriate levels of the independent variables that constitute the interaction. Now consider the other interaction column, the female by associate professor column (column 6). This is equal to the product of the female column (column 2) and the associate professor column (column 4).

Now that we know what interaction is and how to code it into the model, let us examine our data for the presence of statistically significant interaction. The **X** matrix in Equation (B.19) is the appropriate matrix to represent the inclusion of the interaction terms in the model. We will skip the intermediate steps in finding **b**.

$$\mathbf{b}' = [2105.56 \quad -50.00 \quad -555.56 \quad -72.22 \quad 166.67 \quad -50.00]$$

Examine the differential female by assistant professor interaction parameter. Its value is \$166.67, which means that the combination of a female and assistant professor has a salary almost \$200 higher than one would expect when considering the effects of sex and faculty rank separately.

To test the hypothesis of no interaction

$$H_0: \quad \beta_4 = 0$$
$$\beta_5 = 0$$

the **C** matrix is

$$\mathbf{C} = \begin{bmatrix} 0 & 0 & 0 & 0 & 1 & 0 \\ 0 & 0 & 0 & 0 & 0 & 1 \end{bmatrix}$$

If we perform this test at the 0.05 level, the critical F value is $F_{2,12,0.95} = 3.89$. Our F ratio is 7.46, which is greater than 3.89, and therefore we reject the hypothesis of no interaction.

We must now examine the cell means to determine what type of interaction is present. The cell means are:

Faculty Rank	SEX	
	Female	Male
Assistant professor	\$1667	\$1433
Associate professor	\$1933	\$2133
Full professor	\$2567	\$2900

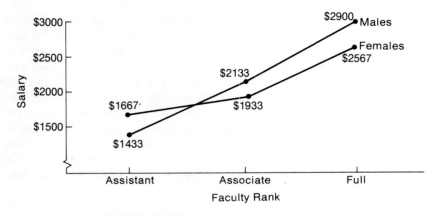

Figure B.2   Observed Sex by Faculty Rank Interaction

The interaction here appears to be the degree type shown in Figure B.2, where female assistant professors have a higher salary than male assistant professors. However, the male associate professors and full professors have higher salaries than the female faculty members at these ranks. It is not possible to talk about a sex effect for these data because females have higher mean salaries for one rank and lower mean salaries for the others. It is possible to have a main effect for faculty rank here, however, because the sex variable does not affect the ordering of faculty ranks. Assistant professors have the lowest mean salaries; the mean of the associate professors is in the middle; and the mean of the full professors is the highest regardless of gender.

SUMMARY

This completes our discussion of the ANOVA procedure. We have not attempted to cover every possible topic; for example, we have not explained how to assess the fit of the model or how to examine residuals. Our discussion has focused on the different ways of coding or reparameterizing the model, the concept and coding of interaction, and the matrix form of calculating the estimates of β and the test statistics.

These are the key topics discussed in this appendix:

- *Traditional ANOVA:* A review of the between-group and within-group sums of squares approach.
- *Linear model ANOVA:* Use of $\mathbf{Y} = \mathbf{X}\boldsymbol{\beta} + \mathbf{e}$ to describe the model that is the foundation of ANOVA.
- *Coding methods:* Three methods were examined; they affect the interpretations of the parameters but not the tests of hypotheses.

- *Tests of hypotheses:* Hypotheses are expressed as H_0: $\mathbf{C\beta} = 0$.
- *Two-way ANOVA:* The problem becomes larger but the same equations—$\mathbf{Y} = \mathbf{X\beta} + \mathbf{e}$ and $\mathbf{b} = (\mathbf{X'X})^{-1}\mathbf{X'Y}$—still apply.
- *Interaction:* The main effects do not explain all the variation in the data and additional terms are required; there are two types of interaction: crossover and degree.

We have emphasized these topics because they form the basis for the analyses presented in the book.

EXERCISES

B.1.

Consider the data in the following table:

		Factor S		
		1	2	3
Factor R	1	10, 15	16, 12	18, 14
	2	15, 11	13, 16	13, 19

The data represent two observations for each of the six combinations of the levels of factors R and S.

a. Construct the appropriate design ($\mathbf{X}$) matrix that represents the relationship between the observed data and the main effects of both factors (R and S) as well as the effect of the general mean term. The effects are to be measured from their mean effects (differential effects coding).

b. Construct the appropriate $\mathbf{X}$ matrix when the effects are measured from one of the levels for both factors R and S (dummy variable coding).

c. The $\mathbf{\beta}$ vector is

$$\mathbf{\beta} = \begin{bmatrix} \beta_0 \\ \beta_1 \\ \beta_2 \\ \beta_3 \end{bmatrix}$$

where β_0 = mean

β_1 = differential effect of level 1 of R

β_2 = differential effect of level 1 of S

β_3 = differential effect of level 2 of S

If we wish to test the hypothesis of no effect of factor S, how can we express this hypothesis? What is the appropriate $\mathbf{C}$ matrix for this hypothesis when it is expressed as H_0: $\quad \mathbf{C}\beta = \mathbf{0}$?

d. Let $\mathbf{X} = \mathbf{I}$ and solve for β. Interpret your answer. Using the hypothesis H_0: $\quad \mathbf{C}\beta = \mathbf{0}$, define a matrix $\mathbf{C}$ to test the hypotheses that factor S has no effect.

B.2.

Provide the appropriate $\mathbf{X}$ matrix for the main effects of sex and faculty rank plus a mean term for the salary data presented in this appendix. Use dummy variable coding.

Appendix C: Table of Chi-Square Values

DF \ α	0.250	0.100	0.050	0.025	0.010	0.005	0.001
1	1.32330	2.70554	3.84146	5.02389	6.63490	7.87944	10.828
2	2.77259	4.60517	5.99147	7.37776	9.21034	10.5966	13.816
3	4.10835	6.25139	7.81473	9.34840	11.3449	12.8381	16.266
4	5.38527	7.77944	9.48773	11.1433	13.2767	14.8602	18.467
5	6.62568	9.23635	11.0705	12.8325	15.0863	16.7496	20.515
6	7.84080	10.6446	12.5916	14.4494	16.8119	18.5476	22.458
7	9.03715	12.0170	14.0671	16.0128	18.4753	20.2777	24.322
8	10.2188	13.3616	15.5073	17.5346	20.0902	21.9550	26.125
9	11.3887	14.6837	16.9190	19.0228	21.6660	23.5893	27.877
10	12.5489	15.9871	18.3070	20.4831	23.2093	25.1882	29.588
11	13.7007	17.2750	19.6751	21.9200	24.7250	26.7569	31.264
12	14.8454	18.5494	21.0261	23.3367	26.2170	28.2995	32.909
13	15.9839	19.8119	22.3621	24.7356	27.6883	29.8194	34.528
14	17.1170	21.0642	23.6848	26.1190	29.1413	31.3193	36.123
15	18.2451	22.3072	24.9958	27.4884	30.5779	32.8013	37.697
16	19.3688	23.5418	26.2962	28.8454	31.9999	34.2672	39.252
17	20.4887	24.7690	27.5871	30.1910	33.4087	35.7185	40.790
18	21.6049	25.9894	28.8693	31.5264	34.8053	37.1564	42.312
19	22.7178	27.2036	30.1435	32.8523	36.1908	38.5822	43.820

DF \ α	0.250	0.100	0.050	0.025	0.010	0.005	0.001
20	23.8277	28.4120	31.4104	34.1696	37.5662	39.9968	45.315
21	24.9348	29.6151	32.6705	35.4789	38.9321	41.4010	46.797
22	26.0393	30.8133	33.9244	36.7807	40.2894	42.7956	48.268
23	27.1413	32.0069	35.1725	38.0757	41.6384	44.1813	49.728
24	28.2412	33.1963	36.4151	39.3641	42.9798	45.5585	51.179
25	29.3389	34.3816	37.6525	40.6465	44.3141	46.9278	52.620
26	30.4345	35.5631	38.8852	41.9232	45.6417	48.2899	54.052
27	31.5284	36.7412	40.1133	43.1944	46.9630	49.6449	55.476
28	32.6205	37.9159	41.3372	44.4607	48.2782	50.9933	56.892
29	33.7109	39.0875	42.5569	45.7222	49.5879	52.3356	58.302
30	34.7998	40.2560	43.7729	46.9792	50.8922	53.6720	59.703
40	45.6160	51.8050	55.7585	59.3417	63.6907	66.7659	73.402
50	56.3336	63.1671	67.5048	71.4202	76.1539	79.4900	86.661
60	66.9814	74.3970	79.0819	83.2976	88.3794	91.9517	99.607
70	77.5766	85.5271	90.5312	95.0231	100.425	104.215	112.317
80	88.1303	96.5782	101.879	106.629	112.329	116.321	124.839
90	98.6499	107.565	113.145	118.136	124.116	128.299	137.208
100	109.141	118.498	124.342	129.561	135.807	140.169	149.449

Appendix D:
The GENCAT
Computer Program

This appendix illustrates the use of the GENCAT computer program for performing the analyses discussed in Chapters 4 through 11. Since the GENCAT program (version 1.2) is the principal analytic software developed to perform weighted least squares analysis of multidimensional contingency tables, we discuss the GENCAT program setup. The necessary card image input is given at the end of each chapter. The FUNCAT procedure contained in SAS 79 may also be used in many cases as an alternative to GENCAT.

It is not our purpose here to duplicate the program instructions and documentation of GENCAT or other computer programs. The GENCAT program is described in Landis and others (1976), Stanish and Koch (1978), and Stanish and others (1978). Additional information regarding the acquisition of GENCAT may be obtained by writing:

> Dr. J. Richard Landis
> Department of Biostatistics
> School of Public Health
> University of Michigan
> Ann Arbor, MI 48109

The SAS program is described in Barr, Goodnight, and Sall (1979) and Helwig (1978). Information regarding SAS is available from:

> SAS Institute, Inc.
> P.O. Box 10066
> Raleigh, NC 27605

269

AN OVERVIEW OF GENCAT

GENCAT is a special-purpose computer program that uses weighted least squares estimation in the analysis of multidimensional contingency tables. The program produces an array of statistics, including estimates of the parameters of the linear model, residuals, predicted values of the functions, standard errors, and assorted test statistics.

Figure D.1 summarizes GENCAT's major input and output steps. In addition to requiring certain information about the structure of the problem and various output options, the program requires input of data, definition of the functions, specification of a model, and definition of contrasts. At each stage of computing, GENCAT offers a range of options that permit the user to adapt the program to a particular application. Because the various combinations of these options introduce considerable flexibility, we briefly discuss the commonly used ones as an aid to interpreting the input listings in Chapters 4 to 11.

The choice of software and selection of options depends on the computer environment and the availability of resources. We wish to stress that the computer setups provided in this appendix represent only one possible means of finding a solution. Table D.1 summarizes the program setups illustrated in this book in order to provide a guide in selecting software.

Table D.1   GENCAT Computer Setups

Chapter	Frequency Data Input (Cases 1 or 2)	Direct Input (Case 3)	Raw Data Input (Case 4)
4	X		
5	X		
6	X		
7	X		
8	X		
9	X		
10		X	X
11	X[a]	X	

[a] Used as a preliminary step to estimate F and V_F for direct input.

ENTERING THE DATA TO GENCAT

GENCAT permits four distinct means of data input:

- *Case 1:* frequency data
- *Case 2:* frequency data for large problems

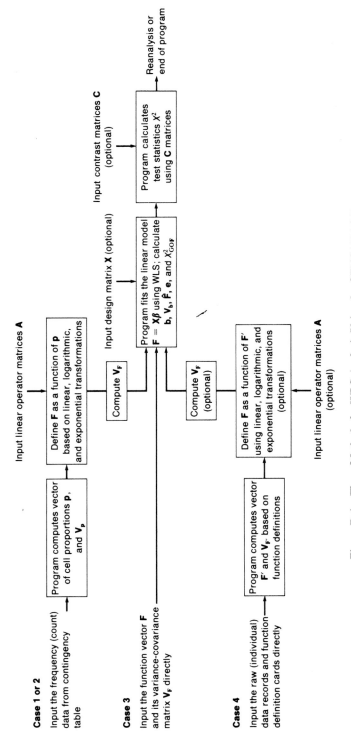

Case 1 or 2

Input the frequency (count) data from contingency table

Input linear operator matrices **A**

Program computes vector of cell proportions **p**, and **V$_p$**

Define **F** as a function of **p** based on linear, logarithmic, and exponential transformations

Compute **V$_F$**

Case 3

Input the function vector **F** and its variance-covariance matrix **V$_F$** directly

Input design matrix **X** (optional)

Program fits the linear model **F** = **Xβ** using WLS; calculate **b**, **V$_b$**, **F̂**, **e**, and X^2_{GOF}

Compute **V$_F$** (optional)

Input contrast matrices **C** (optional)

Program calculates test statistics X^2 using **C** matrices

Reanalysis or end of program

Case 4

Input the raw (individual) data records and function definition cards directly

Program computes vector **F'** and **V$_{F'}$** based on function definitions

Define **F** as a function of **F'** using linear, logarithmic, and exponential transformations (optional)

Input linear operator matrices **A** (optional)

Figure D.1   Three Methods for WLS Analysis Using the GENCAT 1.2 Program

- *Case 3:* direct input of $\mathbf{F}$ and $\mathbf{V}_F$
- *Case 4:* raw data input

The numbering of methods corresponds to the definitions given in the GENCAT 1.2 program input on the first parameter card. (See the program write-up for further details.)

   Close examination of Figure D.1 shows that GENCAT produces the relevant statistical information after $\mathbf{F}$ and $\mathbf{V}_F$ are supplied. For case 1, case 2, and case 4, it is usually necessary to define the left-hand side of the equation $\mathbf{F} = \mathbf{X}\boldsymbol{\beta}$, where $\mathbf{F}$ is the function vector. GENCAT then computes $\mathbf{F}$ and $\mathbf{V}_F$. For case 3 (direct input) this step is accomplished by using software external to GENCAT. For large problems the direct method of input may be the most efficient, since the information contained in $\mathbf{F}$ and $\mathbf{V}_F$ is reduced before undertaking a weighted least squares analysis.

LEFT-HAND SIDE OF THE EQUATION

The vector of functions $\mathbf{F}$, the left-hand side of the equation, contains the information to which we fit the linear model $\mathbf{X}\boldsymbol{\beta}$. When case 1, case 2 (frequency data), or case 4 (raw data) input is used, $\mathbf{F}$ is defined by the input of a linear operator matrix $\mathbf{A}$; by the selection of logarithmic or exponential transformations of the information; or by some combination of the three. GENCAT transforms the frequency information into $\mathbf{F}$ by manipulating the input arrays using matrix operations specified by the user. Whenever the user requires a linear operation $\mathbf{A}$, it must be input directly to GENCAT. Logarithmic and exponential matrix operations are generated directly by GENCAT on command of the user. For most analyses the user will input at least one $\mathbf{A}$ matrix—for example, perform one linear operation—on the input information.

RIGHT-HAND SIDE OF THE EQUATION

Once the user has defined the left-hand side of the equation, the design matrix $\mathbf{X}$ (on the right-hand side) must be defined. This is accomplished by the direct input of the matrix $\mathbf{X}$. This matrix defines the parameters β_i to be estimated using weighted least squares. Once this has been done, GENCAT will compute the estimates $\mathbf{b}_i$, standard errors, a goodness-of-fit X^2 (for nonsaturated models), predicted values, and residuals.

TESTING INDIVIDUAL HYPOTHESES

Tests of hypotheses concerning the population parameters are performed by computing linear transformations of the estimated parameter vector $\mathbf{b}$. The form of the test is $H_0: \quad \mathbf{C}\boldsymbol{\beta} = \mathbf{0}$. The matrix $\mathbf{C}$ is input by the user. The user may input successive $\mathbf{C}$ matrices to perform multiple tests.

SUMMARY OF MAJOR INPUT TO GENCAT

The major input to GENCAT is as follows:

- *Input of data:* The input of frequency data (cases 1 or 2), direct input of **F** and $\mathbf{V}_F$ (case 3), or raw data (case 4).
- *Input of transformations:* The definition of **F** using linear, logarithmic, or exponential operations. This step is usually omitted when case 3 input is used. When linear operations are used, the matrix **A** must be input.
- *Input of the design matrix:* The design matrix **X** must be input by the user. This matrix defines the right-hand side ("linear model" side) of the equation.
- *Input of tests of hypotheses:* The user must input a matrix **C** for each test performed. This step may be repeated as needed.

GENCAT INPUT AND OUTPUT
FOR CHAPTER 4

This section summarizes the major input and output steps of the GENCAT 1.2 program in performing the analysis of data discussed in Chapter 4. The exact control card input is listed at the end of Chapter 4.

Input of Data

The frequency data reported in Table 4.1 are input using the case 1 option. The program then computes the proportion in each cell, subject to the constraint that the proportions in each row sum to 1.00 (see below). These proportions form the estimated probability vector **p**.

```
CONTINGENCY TABLE:

      38.           17.

     244.           21.

      67.           42.

     302.           67.

PROBABILITY VECTOR:

  0.69091D+00      0.30909D+00

  0.92075D+00      0.79245D-01

  0.61468D+00      0.38532D+00

  0.81843D+00      0.18157D+00
```

Input of Transformations

The definition of f_i, the proportion receiving a prison sentence, requires the use of linear transformation. The general form of the linear transformation is

$$\underset{(d \times 1)}{\mathbf{F}} = \underset{(d \times rs)}{\mathbf{A}} \quad \underset{(rs \times 1)}{\mathbf{p}}$$

For this problem **A** is of size (4 × 8):

$$\underset{(4 \times 8)}{\mathbf{A}} = \begin{bmatrix} 0 & 1 & 0 & 0 & 0 & 0 & 0 & 0 \\ 0 & 0 & 0 & 1 & 0 & 0 & 0 & 0 \\ 0 & 0 & 0 & 0 & 0 & 1 & 0 & 0 \\ 0 & 0 & 0 & 0 & 0 & 0 & 0 & 1 \end{bmatrix}$$

Many **A** matrices have a block-diagonal form. This matrix **A** has such a form, where the submatrix $\mathbf{A^*} = \begin{bmatrix} 0 & 1 \end{bmatrix}$ appears in the diagonal of **A**. The relationship of **A*** to **A** is

$$\underset{(4 \times 8)}{\mathbf{A}} = \begin{bmatrix} \mathbf{A^*} & & & \\ & \mathbf{A^*} & & \\ & & \mathbf{A^*} & \\ & & & \mathbf{A^*} \end{bmatrix}$$

When **A** has a block-diagonal form, the GENCAT program requires only the input of **A***. Observe that the operator **A** for this problem selects the second, fourth, sixth, and eighth entries of the vector **p**, which are the proportions receiving a prison sentence for each of the four subpopulations. We have now defined a suitable function vector **F**. Once the function vector **F** has been defined, the program will compute the estimated variance-covariance matrix of **F**. This matrix is designated $\mathbf{V}_F$. The computer output for **A***, **F**, and $\mathbf{V}_F$ is

```
BASIC BLOCK OF LINEAR OPERATOR A1:

   0.0      1.00

F(P) = A1*P:

   0.30909D+00     0.79245D-01     0.38532D+00     0.18157D+00


COVARIANCE MATRIX:

   0.38828D-02     0.0             0.0             0.0

   0.0             0.27534D-03     0.0             0.0

   0.0             0.0             0.21729D-02     0.0

   0.0             0.0             0.0             0.40272D-03
```

Input of the Design Matrix

The design matrix **X** defines the parameters to be estimated. The GENCAT program requires the input of the rank of **X** (the number of columns of the **X** matrix) and the **X** matrix. For this analysis, the rank of **X** is 3 and the matrix **X** is

```
DESIGN MATRIX:      MAIN EFFECTS MODEL

   1.00      1.00      1.00

   1.00      1.00     -1.00

   1.00     -1.00      1.00

   1.00     -1.00     -1.00
```

Once the design matrix has been entered, GENCAT then computes the weighted least squares estimate **b** of β:

```
ESTIMATED MODEL PARAMETERS:

   0.23703D+00    -0.49849D-01    0.10687D+00
```

the estimated variance-covariance matrix $\mathbf{V_b}$ of the estimated parameters **b**:

```
COVARIANCE MATRIX:

   0.38953D-03    0.72031D-05    0.30912D-03

   0.72031D-05    0.15245D-03    0.35841D-04

   0.30912D-03    0.35841D-04    0.39762D-03
```

the estimated standard errors of **b**:

```
STANDARD DEVIATIONS OF THE ESTIMATED MODEL PARAMETERS:

   0.19737D-01    0.12347D-01    0.19940D-01
```

the goodness-of-fit statistic:

```
SQUARE DUE TO ERROR =      0.1011          DF =   1         P =0.7505
```

the observed functions **F**, the expected functions $\hat{\mathbf{F}}$ based on the model $\hat{\mathbf{F}} = \mathbf{Xb}$, the residuals **e**, and the estimated standard errors of the predicted function $\hat{\mathbf{F}}$:

```
F(P) = A1*P:

   0.30909D+00    0.79245D-01    0.38532D+00    0.18157D+00
```

F(P) PREDICTED FROM MODEL:

   0.29404D+00   0.80312D-01   0.39374D+00   0.18001D+00

RESIDUAL VECTOR: (F(P) - PREDICTED F(P))

   0.15048D-01   -0.10671D-02   -0.84210D-02   0.15607D-02

STANDARD DEVIATIONS OF THE PREDICTED FUNCTIONS:

   0.40545D-01   0.16251D-01   0.38363D-01   0.19459D-01

Input of Tests of Hypotheses

After evaluating the model to determine whether the fit is acceptable, one can test specific hypotheses under the model. The two hypotheses tested in this analysis are whether β_1 or β_2 is equal to zero. Hypotheses are tested by entering a suitable matrix C_i, which defines the desired hypothesis. The general form of the hypothesis is H_0: $C\beta = 0$.

To test the hypothesis that $\beta_1 = 0$, the matrix is $C_1 = [0 \quad 1 \quad 0]$. The resulting GENCAT output is

CONTRAST MATRIX: TEST: B1 = 0

   0.0   1.00   0.0

ESTIMATED MODEL CONTRASTS:

  -0.49849D-01

STANDARD DEVIATIONS OF THE ESTIMATED MODEL CONTRASTS:

  0.12347D-01

CHI-SQUARE =   16.3006   DF =   1   P = 0.0001

The test of the hypothesis $\beta_2 = 0$ is carried out by entering the matrix $C_2 = [0 \quad 0 \quad 1]$, and the resulting output is

CONTRAST MATRIX: TEST: B2 = 0

   0.0   0.0   1.00

ESTIMATED MODEL CONTRASTS:

  0.10687D+00

STANDARD DEVIATIONS OF THE ESTIMATED MODEL CONTRASTS:

0.19940D-01

CHI-SQUARE =   28.7217   DF =   1   P = -.0000

This discussion has covered the most important parts of the GENCAT program output used in Chapter 4. Analyses in subsequent chapters, though longer and more complicated, will produce output similar to the parts described here.

References

ANDERSEN, R. and O. W. ANDERSON (1967)
A Decade of Health Services. Chicago, IL: University of Chicago Press.

ANDERSON, T. R. and M. ZELDITCH, Jr. (1968)
A Basic Course in Statistics with Sociological Applications. 2nd ed. New York: Holt, Rinehart, and Winston.

ANSCOMBE, F. J. (1961)
"Examination of Residuals." *Proceedings of the Fourth Berkeley Symposium on Mathematical Statistics and Probability* 1:1–36.

AVNET, H. H. (1967)
Physical Service Patterns and Illness Rates. New York: Group Health Insurance.

BAHR, H. M. (1969)
"Institutional Life, Drinking, and Disaffiliation." *Social Problems* 16:365–376.

BARR, A. J., J. H. GOODNIGHT, and J. P. SALL (1979)
A User's Guide to SAS 79. Raleigh, NC: SAS Institute.

BEAVER, R. J. (1977)
"Weighted Least Squares Response Surface Fitting in Factorial Paired Comparisons." *Communications in Statistics*, A6:1275–1288.

BERKSON, J. (1944)
"Application of the Logistic Function to Bioassay." *Journal of the American Statistical Association* 39:357–365.

BERKSON, J. (1946)
"Approximation of Chi-Square by 'Probits' and by 'Logits.'" *Journal of the American Statistical Association* 41:70–74.

BERKSON, J. (1953)
"A Statistically Precise and Relatively Simple Method of Estimating the Bioassay with Quantal Response Based on the Logistic Function." *Journal of the American Statistical Association* 48:565–599.

BERKSON, J. (1955)
"Maximum Likelihood and Minimum χ^2 Estimates of the Logistic Function." *Journal of the American Statistical Association* 50:130–162.

BERKSON, J. (1968)
"Application of Minimum Logit χ^2 Estimate to a Problem of Grizzle with a Notation on the Problem of 'No Interaction.'" *Biometrics* 24:75–95.

BERKSON, J. (1972)
"Minimum Discrimination Information, the 'No Interaction' Problem, and the Logistic Function." *Biometrics* 28:443–468.

BHAPKAR, V. P. (1966)
"A Note on the Equivalence of the Two Test Criteria for Hypothesis in Categorical Data." *Journal of the American Statistical Association* 61:228–235.

BHAPKAR, V. P. (1979)
"On Bias Reduction of the Wald Statistic for Testing Hypotheses in Categorical Data." University of Kentucky, Department of Statistics, Technical Report No. 139.

BHAPKAR, V. P. and G. G. KOCH (1968a)
"Hypothesis of 'No Interaction' in Multi-Dimensional Contingency Tables." *Technometrics* 10:107–123.

BHAPKAR, V. P. and G. G. KOCH (1968b)
"On the Hypothesis of 'No Interaction' in Contingency Tables." *Biometrics* 24:567–594.

BISHOP, Y.M.M., S. E. FIENBERG, and P. W. HOLLAND (1975)
Discrete Multivariate Analysis: Theory and Practice. Cambridge, MA: MIT Press.

BOX, G. E. P., W. D. HUNTER, and J. S. HUNTER (1978)
Statistics for Experimenters: An Introduction to Design, Data Analysis, and Model Building. New York: John Wiley and Sons.

BRADLEY, R. A. and M. E. TERRY (1952)
"Rank Analysis of Incomplete Block Designs. I: The Method of Paired Comparisons." *Biometrika* 39:324–345.

BRESLOW, N. E. and N. E. DAY (1975)
"Indirect Standardization and Multiplicative Models for Rates, with Reference to the Age Adjustment of Cancer Incidence and Relative Frequency Data." *Journal of Chronic Disease* 28:289–303.

CHEN, T. and S. E. FIENBERG (1976)
"The Analysis of Contingency Tables with Incomplete Classified Data." *Biometrics* 32:123–144.

CHIANG, C. L. (1968)
Introduction to Stochastic Process in Biostatistics. New York: John Wiley and Sons.

CLARKE, S. H. and G. G. KOCH (1976)
"The Influence of Income and Other Factors on Whether Criminal Defendants Go to Prison." *Law and Society Review* 11:57–92.

CLARKE, S. H., J. L. FREEMAN, and G. G. KOCH (1976)
"Bail Risk: A Multivariate Analysis." *Journal of Legal Studies* 5: 341–385.

COCHRAN, W. G. (1954)
"Some Methods for Strengthening the Common χ^2 Tests." *Biometrics* 10:417–451.

CONVERSE, P. E. (1964)
"The Nature of Belief Systems in Mass Publics." In D. E. Apter (ed.), *Ideology and Discontent*. New York: Free Press.

CORN, R. F., J. R. LANDIS, and J. D. FLORA (1977)
"A Comparative Evaluation of Two Roadside Brake Testing Procedures." *Accident Analysis and Prevention* 9:167–176.

CORNFIELD, J. (1967)
"Discriminant Functions." *International Statistical Review* 35:142–153.

COX, D. R. (1970)
The Analysis of Binary Data. London: Methuen.

COX, D. R. (1972a)
"The Analysis of Multivariable Binary Data." *Applied Statistics* 21:113–120.

COX, D. R. (1972b)
"Regression Models and Life Tables (with Discussion)." *Journal of the Royal Statistical Society*, Series B, 34:187–220.

COX, D. R. and E. J. SNELL (1968)
"A General Definition of Residuals." *Journal of the Royal Statistical Society*, Series B, 30:248–265.

CUTLER, S. J. and F. EDERER (1958)
"Maximum Utilization of the Life Table Method in Analyzing Survival." *Journal of Chronic Disease* 8:699–712.

CUTLER, S. J. and M. H. MYERS (1967)
"Clinical Classification of Extent of Disease in Cancer of the Breast." *Journal of the National Institute of Cancer* 39:193–207.

DRAPER, N. and H. SMITH (1966)
Applied Regression Analysis. New York: John Wiley and Sons.

DYKE, G. U. and H. D. PATTERSON (1952)
"Analysis of Factorial Arrangements When Data Are Proportions." *Biometrics* 8:1–12.

EL-KHORAZATY, M. N. and others (1976)
"On Log-Linear Models for Multiple-Record Systems." *Proceedings of the American Statistical Association, Social Statistics Section.*

EL-KHORAZATY, M. N. and others (1977)
"Estimating the Total Number of Events with Data from Multiple-Record Systems: A Review of Methodological Strategies." *International Statistical Review* 45:129–157.

EVERITT, B. S. (1977)
The Analysis of Contingency Tables. London: Chapman and Hall.

FIENBERG. S. E. (1977)
The Analysis of Cross-Classified Categorical Data. Cambridge, MA: MIT Press.

FINNEY, D. J. (1964)
Statistical Method in Biological Assay. 2nd ed. London: Griffin.

FLEISS, J. L. (1973)
Statistical Methods for Rates and Proportions. New York: John Wiley and Sons.

FORTHOFER, R. N. and G. G. KOCH (1973)
"An Analysis for Compounded Functions of Categorical Data." *Biometrics* 2:143–157.

FORTHOFER, R. N. and G. G. KOCH (1974)
"A Program for the Analysis for Compounded Functions of Categorical Data." *Computer Programs in Biomedicine* 3:237–248.

FORTHOFER, R., J. GLASSER, and N. LIGHT (1980)
"Life Table Analysis of Membership Retention in an H.M.O." *Journal of Community Health* 5:46–53.

FORTHOFER, R. N., C. F. STARMER, and J. E. GRIZZLE (1971)
"A Program for the Analysis of Categorical Data by Linear Models." *Journal of Biomedical Systems* 2:3–48.

FREEMAN, D. H. Jr., J. L. FREEMAN, and G. G. KOCH (1978)
"A Modified Chi-Square Approach for Fitting Weibull Models to Synthetic Life Tables." *Biometrische Zeitschrift* 20:29–40.

FREEMAN, J. L. and others (1975)
"Shoulder Harness Usage in the Population of Drivers at Risk in North Carolina." Chapel Hill: Highway Safety Research Center, University of North Carolina.

FREEMAN, D. H. Jr. and others (1976a)
"Strategies in the Multivariate Analysis of Data from Complex Surveys. II: An Application to the United States National Health Interview Survey." *International Statistical Review* 44:317–330.

FREEMAN, D. H. Jr. and others (1976b)
"An Analysis of Physician Visit Data from a Complex Sample Survey." *American Journal of Public Health* 66:979–983.

FROHARDT-LANE, K. A., J. R. LANDIS, and W. H. BRUVOLD (1977)
"A New Technique for Measuring Preferences in Demographic Studies." *Demography* 14:97–102.

GEHAN, E. A. and M. M. SIDDIQUI (1973)
"Simple Regression Methods for Survival Time Studies." *Journal of the American Statistical Association* 68:848–856.

GOKHALE, D. V. and S. KULLBACK (1978)
The Information in Contingency Tables. New York: M. Dekker.

GOODMAN, L. A. (1970)
"The Multivariate Analysis of Qualitative Data: Interaction Among Multiple Classifications." *Journal of the American Statistical Association* 65:226–256.

GOODMAN, L. A. (1971)
"The Partitioning of Chi-Square, the Analysis of Marginal Contingency Tables, and the Estimation of Expected Frequencies in Multidimensional Contingency Tables." *Journal of the American Statistical Association* 66:339–344.

GOODMAN, L. A. (1972)
"A Modified Multiple Regression Approach to the Analysis of Dichotomous Variables." *American Sociological Review* 37:28–46.

GOODMAN, L. A. (1973)
"Causal Analysis of Data from Panel Studies and Other Kinds of Surveys." *American Journal of Sociology* 78:1135–1191.

GOODMAN, L. A. and W. H. KRUSKAL (1954)
"Measures of Association for Cross-Classifications." *Journal of the American Statistical Association* 49:732–764.

GOODMAN, L. A. and W. H. KRUSKAL (1959)
"Measures of Association for Cross-Classifications: Further Discussions and References." *Journal of the American Statistical Association* 54:123–163.

GOODMAN, L. A. and J. MAGIDSON, editors (1978)
Analyzing Qualitative/Categorical Data: Log-Linear Models and Latent-Structure Analysis. Cambridge, MA: Abt Books.

GRAYBILL, F. A. (1969)
Introduction to Matrices with Applications in Statistics. Belmont, CA: Wadsworth Publishing Co.

GREENLICK, M. R. and others (1968)
"Determinants of Medical Care Utilization." *Health Services Research* 3:296–315.

GRIZZLE, J. E., C. F. STARMER, and G. G. KOCH (1969)
"Analysis of Categorical Data for Linear Models." *Biometrics* 25:489–504.

HABERMAN, S. J. (1978)
Analysis of Qualitative Data. New York: Academic Press.

HALPERIN, M., W. C. BLACKWELDER, and J. I. VERTER (1971)
"Estimation of the Multivariate Logistic Risk Function: A Comparison of the Discriminant Function and Maximum Likelihood Approach." *Journal of Chronic Disease* 24:125–158.

HELWIG, J. T. (1978)
   SAS Introductory Guide. Raleigh, NC: SAS Institute.

HIGGINS, J. E. and G. G. KOCH (1977)
   "Variable Selection and Generalized Chi-Square Analysis of Categorical Data Applied to a Large Cross-Sectional Occupational Health Survey." *International Statistical Review* 45:51–62.

HILDEBRAND, D. K., J. D. LAING, and H. ROSENTHAL (1977)
   Analysis of Ordinal Data. Beverly Hills, CA: Sage Publications.

HOEL, G. (1971)
   Elementary Statistics. 3rd ed. New York: John Wiley and Sons.

IMREY, P. B. and G. G. KOCH (1972)
   "Linear Models Analysis of Incomplete Multi-Variate Categorical Data." University of North Carolina, Institute of Statistics, *Mimeo Series*, No. 820.

IMREY, P. B., W. D. JOHNSON, and G. G. KOCH (1976)
   "An Incomplete Contingency Table Approach to Paired-Comparison Experiments." *Journal of the American Statistical Association* 71:614–623.

IMREY, P. B., E. SOBEL, and M. E. FRANCIS (1979)
   "Analysis of Categorical Data Obtained by Stratified Random Sampling." *Communications in Statistics*, A8:653–670.

JOHNSON, W. D. and G. G. KOCH (1970)
   "Analysis of Qualitative Data: Linear Functions." *Health Services Research* 5:358–369.

JOHNSON, W. D. and G. G. KOCH (1971)
   "A Note on the Weighted Least-Squares Analysis of the Ties–Smith Contingency Table Data." *Technometrics* 13:438–447.

JOHNSON, W. D. and G. G. KOCH (1978)
   "Linear Models Analysis of Competing Risks for Grouped Survival Times." *International Statistical Review* 46:21–51.

KENDALL, M. G. (1955)
   Rank Correlation Methods. 2nd ed. London: Griffin.

KERSHNER, R. P. and G. C. CHAO (1976)
   "A Comparison of Some Categorical Analysis Programs." *Proceedings of the American Statistical Association, Statistical Computing Section, Boston.*

KLEINBAUM, D. and L. KUPPER (1978)
   Applied Regression Analysis and Other Multivariable Methods. North Scituate, MA: Duxbury Press.

KOCH, G. G. (1968)
   "The Effect of Non-Sampling Errors on Measures of Association in 2 × 2 Contingency Tables." *Journal of the American Statistical Association* 63:852–863.

KOCH, G. G. (1969)
"Some Aspects of the Statistical Analysis of 'Split Plot' Experiments in Completely Randomized Layouts." *Journal of the American Statistical Association* 64:485–505.

KOCH, G. G. (1973)
"An Alternative Approach to Multivariate Response Error Models for Sample Survey Data with Applications to Estimators Involving Subclass Means." *Journal of the American Statistical Association* 68:906–913.

KOCH, G. G. and V. P. BHAPKAR (1981)
"Chi-Square Tests." In N. L. Johnson and S. Kotz (eds.), *Encyclopedia of Statistical Sciences*. New York: John Wiley and Sons, 1981.

KOCH, G. G. and D. W. REINFURT (1971)
"The Analysis of Categorical Data from Mixed Models." *Biometrics* 27:157–173.

KOCH, G. G. and D. W. REINFURT (1974)
"An Analysis of the Relationship Between Driver Injury and Vehicle Age for Automobiles Involved in North Carolina Accidents During 1966–1970." *Accident Analysis and Prevention* 6:1–18.

KOCH, G. G. and M. E. STOKES (1981)
"Chi-Square Tests: Numerical Examples." In N. L. Johnson and S. Kotz (eds.), *Encyclopedia of Statistical Sciences*. New York: John Wiley and Sons, 1981.

KOCH, G. G. and H. D. TOLLEY (1975)
"A Generalized Modified χ^2 Analysis of Categorical Bacteria Survival Data from a Complex Dilution Experiment." *Biometrics* 31:59–12.

KOCH, G. G., J. R. ABERNATHY, and P. B. IMREY (1975)
"On a Method for Studying Family Size Preferences." *Demography* 12:57–66.

KOCH, G. G., M. N. EL-KHORAZATY, and A. L. LEWIS (1976)
"The Asymptotic Covariance Structure of Log-Linear Model Estimated Parameters for the Multiple Recapture Census." *Communications in Statistics* 15:1425–1455.

KOCH, G. G., D. H. FREEMAN Jr., and J. L. FREEMAN (1975)
"Strategies in the Multivariate Analysis of Data from Complex Surveys." *International Statistical Review* 43:59–78.

KOCH, G. G., J. L. FREEMAN, and R. G. LEHNEN (1976)
"A General Methodology for the Analysis of Ranked Policy Preference Data." *International Statistical Review* 44:1–28.

KOCH, G. G., D. B. GILLINGS, and M. E. STOKES (1980)
"Biostatistical Implications of Design, Sampling, and Measurement to Health Science Data Analysis." *Annual Review of Public Health* 1:163–225.

KOCH, G. G., P. B. IMREY, and D. W. REINFURT (1972)
"Linear Model Analysis of Categorical Data with Incomplete Response Vectors." *Biometrics* 28:633–692.

KOCH, G. G., W. D. JOHNSON, and H. D. TOLLEY (1972)
"A Linear Models Approach to the Analysis of Survival and Extent of Disease in Multidimensional Contingency Tables." *Journal of the American Statistical Association* 67:783–796.

KOCH, G. G., H. D. TOLLEY, and J. L. FREEMAN (1976)
"An Application of the Clumped Binomial Model to the Analysis of Clustered Attribute Data." *Biometrics* 32:337–354.

KOCH, G. G. and others (1976)
"The Asymptotic Covariance Structure of Estimated Parameters from Contingency Table Log-Linear Models." *Proceedings of the Ninth International Biometric Conference, Boston.*

KOCH, G. G. and others (1977)
"A General Methodology for the Analysis of Experiments with Repeated Measurement of Categorical Data." *Biometrics* 33:133–158.

KOCH, G. G. and others (1978)
"Symposium: Multivariate and Discrete Data Analysis in Animal Populations: Statistical Methods for Evaluation of Mastitis Treatment Data." *Journal of Dairy Science* 61:830–847.

KU, H. H. and S. KULLBACK (1974)
"Log-Linear Models in Contingency Table Analysis." *American Statistician* 28:115–122.

KUECHLER, M. (1980)
"The Analysis of Nonmetric Data." *Sociological Methods and Research* 8:369–388.

LAIRSON, D. R., R. N. FORTHOFER, and J. H. GLASSER (1979)
"Catastrophic Illness in an HMO." *Inquiry* 16:119–130.

LANDIS, J. R. and G. G. KOCH (1974)
"A Review of Statistical Methods in the Analysis of Data Arising from Observer Reliability Studies." *Statistica Neerlandica* 3, pt. I:101–123; pt. II:151–161.

LANDIS, J. R. and G. G. KOCH (1977a)
"An Application of Hierarchical Kappa Type Statistics in the Assessment of Majority Agreement Among Multiple Observers." *Biometrics* 33:363–374.

LANDIS, J. R. and G. G. KOCH (1977b)
"The Measurement of Observer Agreement for Categorical Data." *Biometrics* 33:159–174.

LANDIS, J. R. and G. G. KOCH (1977c)
"A One-Way Components of Variance Model for Categorical Data." *Biometrics* 33:671–679.

LANDIS, J. R., E. R. HEYMAN, and G. G. KOCH (1978)
"Average Partial Association in Three-Way Contingency Tables: A Review and Discussion of Alternative Tests." *International Statistical Review* 46:237–254.

LANDIS, J. R. and others (1976)
"A Computer Program for the Generalized Chi-Square Analysis of Categorical Data Using Weighted Least Squares (GENCAT)." *Computer Programs in Biomedicine* 6:196–231.

LARNTZ, K. (1978)
"Small Sample Comparisons of Exact Levels for Chi-Squared Goodness of Fit Statistics." *Journal of the American Statistical Association* 73:253–263.

LAZARSFELD, P. F. (1948)
"The Use of Panels in Social Research." *Proceedings of the American Philosophical Society* 92:405–510.

LEHNEN, R. G. (1976)
American Institutions, Political Opinion and Public Policy. Hinsdale, IL: Dryden Press.

LEHNEN, R. G. and G. G. KOCH (1973)
"A Comparison of Conventional and Categorical Regression Techniques in Political Analysis." *Proceedings of the American Political Science Association, New Orleans.*

LEHNEN, R. G. and G. G. KOCH (1974a)
"The Analysis of Categorical Data from Repeated Measurement Research Designs." *Political Methodology* 1:103–123.

LEHNEN, R. G. and G. G. KOCH (1974b)
"Analyzing Panel Data with Uncontrolled Attrition." *Public Opinion Quarterly* 38:40–56.

LEHNEN, R. G. and G. G. KOCH (1974c)
"A General Linear Approach to the Analysis of Non-Metric Data: Applications for Political Science." *American Journal of Political Science* 18:283–313.

LEHNEN, R. G. and A. J. REISS, Jr. (1978)
"Response Effects in the National Crime Survey." *Victimology* 3:110–124.

LEWONTIN, R. C. and J. FELSENSTEIN (1965)
"The Robustness of Homogeneity Tests in 2 × N Tables." *Biometrics* 21:19–33.

LOMBARD, H. L. and C. R. DOERING (1947)
"Treatment of the Four-Fold Table by Partial Correlation As It Relates to Public Health Problems." *Biometrics* 3:123–128.

MANTEL, N. (1963)
"Chi-Square Tests with One Degree of Freedom: Extension of the Mantel-Haenszel Procedure." *Journal of the American Statistical Association* 58:690–700.

MAXWELL, A. E. (1961)
Analyzing Quantitative Data. London: Methuen.

McCLIMANS, C. D. (1978)
"A Study of the Relationship Between Respiratory Morbidity, Air Pollution, and Occupational Exposure in Southeast Texas." Unpublished dissertation, University of Texas School of Public Health.

MERRELL, M. and L. SHULMAN (1955)
"Determination of Prognosis in Chronic Disease, Illustrated by Systemic Lupus Erythematosus." *Journal of Chronic Disease* 1:12–32.

MORRISON, D. G. (1967)
Multivariate Statistical Methods. New York: McGraw-Hill.

MOSTELLER, F. (1968)
"Association and Estimation in Contingency Tables." *Journal of the American Statistical Association* 63:1–28.

NELDER, J. A. and R.W.M. WEDDERBURN (1972)
"Generalized Linear Models." *Journal of the Royal Statistical Society*, Series A, 135:370–384.

NERLOVE, M. and S. J. PRESS (1973)
Univariate and Multivariate Log-Linear and Logistic Models. Report R-1306-EDA/NIH. Santa Monica, CA: Rand Corporation.

POPE, C. (1978)
"Consumer Satisfaction in a Health Maintenance Organization." *Journal of Health and Social Behavior* 19:291–303.

POTTER, R. G. Jr. (1966)
"Application of Life Table Techniques to Measurement of Contraceptive Effectiveness." *Demography* 3:297–304.

REINFURT, D. W. (1970)
"The Analysis of Categorical Data with Supplemented Margins Including Applications to Mixed Models." University of North Carolina, Institute of Statistics, Mimeo Series No. 697.

SCHEFFÉ, H. (1959)
The Analysis of Variance. New York: John Wiley and Sons.

SCHMIDT, P. and R. P. STRAUSS (1975)
"The Prediction of Occupation Using Multiple Logit Models." *International Economic Review* 16:471–486.

SEARLE, S. R. (1966)
Matrix Algebra for the Biological Sciences. New York: John Wiley and Sons.

SIEGEL, S. (1956)
Nonparametric Statistics for the Behavioral Sciences. New York: McGraw-Hill.

SOMERS, R. H. (1962)
"A New Asymmetric Measure of Association for Ordinal Variables." *American Sociological Review* 27:799–811.

STAFFORD, J. E. and R. G. LEHNEN (1975)
Houston Community Study. Houston: Hearne Publishing Company.

STANISH, W. M. and G. G. KOCH (1978)
"A Computer Program for Multivariate Ratio Analysis (MISCAT)." *Computer Programs in Biomedicine* 8:197–207.

STANISH, W. M. and others (1978)
"A Computer Program for the Generalized Chi-Square Analysis of Computing Risks Group Survival Data (CRISCAT)." *Computer Programs in Biomedicine* 8:208–223.

STEINER, J. and R. G. LEHNEN (1974)
"Political Status and Norms of Decision-Making." *Comparative Political Studies* 7:84–106.

STOUFFER, S. A. and others (1949)
The American Soldier: Adjustment During Army Life. Studies in Social Psychology in World War II, vol. 1. Princeton, NJ: Princeton University Press.

STRANG, G. (1976)
Linear Algebra and Its Applications. New York: Academic Press.

STRUMPF, G. B. and M. A. GARRAMONE (1976)
"Why Some HMOs Develop Slowly." *Public Health Reports* 91:496–503.

THEIL, H. (1970)
"On the Estimation of Relationships Involving Qualitative Variables." *American Journal of Sociology* 76:103–154.

TRUETT, J., J. CORNFIELD, and W. B. KANNEL (1967)
"A Multivariate Analysis of the Risk of Coronary Heart Disease in Framingham." *Journal of Chronic Disease* 20:511–524.

UPTON, G.J.H. (1978)
The Analysis of Cross-Tabulated Data. New York: John Wiley and Sons.

U.S. SENATE, COMMITTEE ON GOVERNMENT OPERATIONS, SUBCOMMITTEE ON INTERGOVERN-MENTAL RELATIONS (1973)
Confidence and Concern: Citizens View American Government. Pts. 1 and 2. Washington, DC: Government Printing Office.

WAGNER, E. H. and others (1976)
"Influence of Training and Experience on Selecting Criteria to Evaluate Medical Care." *New England Journal of Medicine* 294:871–876.

WALKER, S. H. and D. B. DUNCAN (1967)
"Estimation of the Probability of an Event as a Function of Several Independent Variables." *Biometrika* 54:167–179.

WILLIAMS, O. D. and J. E. GRIZZLE (1970)
"Analysis of Categorical Data with More Than One Response Variable by Linear Models." University of North Carolina, Institute of Statistics, Mimeo Series, No. 715.

WOLLSTADT, L. J., S. SHAPIRO, and J. W. BICE (1978)
"Disenrollment from a Prepaid Group Practice: An Actuarial and Demographic Description." *Inquiry* 15:142–150.

YARNOLD, J. K. (1970)
"The Minimum Expectation in χ^2 Goodness of Fit Tests and the Accuracy of Approximations for the Null Distributions." *Journal of the American Statistical Association* 65:864–886.

ZIMMERMAN, H. and V. W. RAHLFS (1976)
"Die Erweiterung des Bradley–Terry–Luce Modells auf Bindungen im Rahmen verallgemeinester (Log-) Linearmodelle." *Biometrische Zeitschrifte* 18:23–32.

ZUBROD, C. G. and others (1960)
"Appraisal of Methods for the Study of Chemotherapy of Cancer in Man: Comparative Therapeutic Trial of Nitrogen Mustard and Triethylene Thiophosphoramide." *Journal of Chronic Disease* 11:7–33.

Index

Analysis of covariance, 9
  logistic covariance, 9
Analysis of variance, 6, 9, 26, 79,
    238–239, 243
  linear models approach, 243
  two-way, 254
  two-way with interaction, 257
Association
  additive, 30, 34, 35
  multiplicative, 31, 34, 38, 102–105

Balanced data, 238
Binomial distribution, 22
Bradley-Terry model, 211

Categorical data analysis, 3, 9
Cochran-Mantel-Haenszel procedure, 213,
    215
Coding methods, 247, 248, 250
  differential effects, 250, 254–255,
      260–261
  dummy, 248–250
Competing risks, 213

Complex functions (*see* Functions,
    complex)
Complex sample surveys, 213–214
Concordant, 151, 153–154
Consistency, 147, 157
Contingency table, 19–20
  multidimensional, 4
  two-way, 3
Continuous choice problem, 165–166

Degrees of freedom, 27, 47, 241, 253
Design matrix (**X**), 27, 29, 43, 45, 244,
    248, 249, 251, 255, 259, 261,272,
    275
Differences, 12, 34
Discordant, 151, 153–154
Discrete choice problem, 165–166, 168

Errors (*see* Residuals)
Error sum of squares, 246

F ratio, 241, 253
Fixed row totals, 21

Forced Expiratory Volume in one second
(FEV$_{1.0}$), 99
Forced Vital Capacity (FVC), 99
FUNCAT, 11, 269
Functions, 4, 9, 10, 11, 22, 272
additive, 34
complex, 11, 22, 24, 31, 156, 179,
205, 213–214
linear, 10, 27
logarithmic, 10, 23
multiple, 121
multiplicative, 30–35, 102–105
multiresponse, 5, 13, 14, 111, 121,
124, 126, 164
response, 43, 44
single-response, 121, 124
Functional Asymptotic Regression
Methodology (FARM), 213–214

Gamma (Goodman and Kruskal), 5, 11,
14, 149–153, 155–156, 158,
160–161
GENCAT, 6, 11, 54, 69, 73, 97, 118,
141, 153, 156–157, 161, 169, 186,
206, 208
data input, 270, 272
direct input, 169
raw data input, 169
Goodness of fit, 11, 17, 27, 29, 47–48,
58, 63, 65, 82–83, 103, 107, 129,
134, 136, 171, 175, 202–203

Health Maintenance Organization (HMO),
6, 20, 23, 30, 75–76, 78, 91–92,
191
Houston Community Study, 5, 165–169,
172, 179
Homoscedasticity, 26
Hypotheses testing, 48, 252, 257, 263,
272, 276
goodness-of-fit, 27, 47, 51
test statistic indifference, 164
linear, 27
marginal homogeneity, 138
test statistics, 27, 29, 48, 51
total indifference, 168, 170–171, 173
total variation, 69

Indifference, hypothesis of, 164, 168–173
Interaction, 47, 55, 57–59, 67, 103–104,
128–129, 202, 263–264
first-order, 58
second-order, 58

Least squares estimation, 246, 256, 263
Life table methodology
follow-up, 191–192, 194, 205, 213
survival, 213
survival curve, 6
survival rate, 194, 206, 208
weighted least squares approach, 194
Linear model (*see* Models, linear)
Linear trend term, 157, 202–204
Log cross product ratio, 115–117
Log linear model (*see* Models, log linear)
Logit, 10–11, 13, 23, 98, 101–106,
109–110 (*see also* Models, log
linear)
l/r rule, 14, 156

Matrix, 219
addition, 221
block diagonal, 220, 221
diagonal, 220
identity, 220, 224, 225
inverse, 226, 227
multiplication, 222, 223, 224
non-singular, 226
scalar multiplication, 222
singular, 226, 247, 260
square, 220
subtraction, 222
symmetric, 221
transpose, 221
Maximum likelihood, 3, 107, 108, 110,
214
Mean, 13
rank score, 166–171
score, 9, 79, 80, 81, 126, 127, 131
transformed rank, 167
Minimum discrimination information, 3
Models
additive, 10
interpreting, 48–49
linear, 3, 24, 26, 30, 35, 44, 238, 243,
245, 254

Models (*continued*)
   log linear, 5, 10, 12, 98
   logistic linear, 103
   main effects, 57–58, 65, 67, 131
   mixed, 212
   modular, 57, 59–61, 65, 67, 202
   multiplicative, 10
   prediction, 81
   saturated, 58, 69
Multinomial distribution, 22, 25
Multiple comparisons, 17
Multiple-record systems, 213
Multi-response analysis, 5, 111,
   121–122, 124, 126, 138, 164, 179
Multivariate analysis of variance
   (MANOVA), 173, 179

Net intergenerational mobility score, 94
Nolle prosequi, 111

Observer agreement, 212
Odds ratio, 108–109

P-value, 16
Paired comparisons, 211
Parameters, 43, 45, 244, 245
Partial association, 213
Pearson chi-square, 215
Philosophy of analysis, 215
   exploratory strategy 15–16
   strict position 15–16
Probability π_{ij}, 20, 22, 43, 124–125
Proportion p_{ij}, 13, 20, 24, 43, 44
   multiple, 131
Prediction equation, 82, 87, 90
Pulmonary Function Test (PFT), 99, 100,
   107–110

Rank choice analysis, 164–179
   continuous choice, 165–166, 172–173,
     179
   discrete choice, 165
     few choices, 166
     many choices, 165–166, 168, 179

Rank correlation coefficient, 5, 10, 24,
   146 (*see also* Gamma)
Ratios, 12, 34 (*see also* Odds ratio)
Regression, 9
   dummy variable, 9
   forward stepwise, 215
   logistic, 9
   parametric, 215
Reparameterization, 247
Repeated measures, 212
Residuals, 46–47, 86, 275–276
   error vector, 245
Retention rate, 196, 201

Sample size
   minimum, 12–14, 76
SAS, 12, 107, 269
Saturated model, 58–61
Scalar, 220
Scale
   additive, 12
   logarithmic, 10
   multiplicative, 12
Significance level, 16, 17
Simple random sample, 22
Social distance, 8
Somers d_{xy} and d_{yx}, 146, 150–151
Subpopulations, 19
Supplemented margins, 212
Survey panel design, 121
Survival life table (*see* Life table
   methodology)
System of linear equations
   matrix presentation, 231
   scalar presentation, 228
   solution by substitution, 228–230

Tau (Kendall), 146, 150–151
Transformation
   exponential, 10, 232–234
   linear, 10, 23, 274
   logarithmic, 10, 23, 232–235

Variables
   categorical, 3, 7

Variables (*continued*)
  continuous, 8
  dependent, 9
  dichotomous, 7
  grouped interval, 7–8
  independent, 9
  metric, 7
  nominal, 7–8
  nonmetric, 7
  nonordered polytomous, 7
  ordered polytomous, 7
  ordinal, 7–8
  qualitative, 7
  quantitative, 7

Variable selection, 241
Variance, 24–25
  between-group estimate, 240, 242
  covariance, 25
  estimated variance, 25–26

Variance (*continued*)
  covariance matrix, 19, 24–27, 45, 51,
    124, 153, 167, 169, 173, 201,
    204, 205
  proportion explained, 69
  within-group estimate, 240, 241, 242
Vector
  column, 220
  row, 220

Weighted least squares, 4, 26, 107–108,
    173, 214
  estimator of β, 26, 45, 51, 237, 275

X_2 matrix (*see* Design matrix)
X statistic (*see* Hypotheses testing)

Zero cells (*see* l/r rule)